Quantum Electronics

Edited by Faustino Wahaia

Published in London, United Kingdom

IntechOpen

Supporting open minds since 2005

Quantum Electronics
http://dx.doi.org/10.5772/intechopen.74447
Edited by Faustino Wahaia

Contributors
Youssef Trabelsi, Gabriel Martinez-Niconoff, Marco Antonio Torres-Rodriguez, Mayra Vargas Morales, Patricia Martinez Vara, Gholamreza Shayeganrad, Leila Mashhadi, Er'El Granot

First published in London, United Kingdom, 2019 by IntechOpen
IntechOpen is the global imprint of INTECHOPEN LIMITED, registered in England and Wales, registration number: 11086078, The Shard, 25th floor, 32 London Bridge Street
London, SE19SG – United Kingdom
Printed in Croatia

British Library Cataloguing-in-Publication Data
A catalogue record for this book is available from the British Library

Additional hard and PDF copies can be obtained from orders@intechopen.com

Quantum Electronics
Edited by Faustino Wahaia
p. cm.
Print ISBN 978-1-83881-215-7
Online ISBN 978-1-83881-216-4
eBook (PDF) ISBN 978-1-83881-217-1

Meet the editor

Faustino Wahaia received his BS and MSc degrees from the University of Sofia St. Kliment Ohridski, Sofia, Bulgaria, and PhD degree in Physics from the University of Porto, Portugal. He did his postdoctoral research focusing on activity in generation, detection, and characterization of terahertz (THz) waves using photoconductive antennas (PCA) at the Institute of Biomedical Engineering (INEB), University of Porto, Portugal, and the State Research Institute of Lithuania Center for Physical Sciences and Technology, Vilnius. Dr. Wahaia's research interests are lasers, ultrafast lasers and optics, Raman spectroscopy, imaging and spectroscopic ellipsometry, infrared and THz photonics for diagnostic, ultrafast biophotonics, nanomaterial studies, food analysis, agriculture, defense, health care, homeland security, environment and pharmacy assessment. He is presently a researcher at Porto University. He is a member of SPIE, OSA Optical Society, IEEE, and EACR.

Contents

Preface **XI**

Chapter 1 **1**
Photonic Quasicrystals for Filtering Application
by Youssef Trabelsi

Chapter 2 **17**
Synthesis of Curved Surface Plasmon Fields through Thin Metal Films in a Tandem Array
by Gabriel Martinez Niconoff, Marco Antonio Torres Rodriguez, Mayra Vargas Morales and Patricia Martinez Vara

Chapter 3 **31**
Localized Excitation of Single Atom to a Rydberg State with Structured Laser Beam for Quantum Information
by Leila Mashhadi and Gholamreza Shayeganrad

Chapter 4 **51**
Single-Atom Field-Effect Transistor
by Er'el Granot

Preface

This Edited Volume is a collection of reviewed and relevant research chapters, concerning the developments within the Quantum Electronics. The book includes scholarly contributions by various authors and edited by a group of experts in the field. Each contribution comes as a separate chapter complete in itself but directly related to the book's topics and objectives. The book includes chapters dealing with the topics: Photonic Quasicrystals for Filtering Application, Synthesis of Curved Surface Plasmon Fields through Thin Metal Films in a Tandem Array, Localized Excitation of Single Atom to a Rydberg State with Structured Laser Beam for Quantum Information and Single-Atom Field-Effect Transistor. The target audience comprises scholars and specialists in the field.

IntechOpen

Chapter 1

Photonic Quasicrystals for Filtering Application

Youssef Trabelsi

Abstract

In this chapter, we study the properties of specific one dimensional photonic quasicrystal (PQCs), in order to design an output multichannel filter. We calculate the transmittance spectrum which exhibits a photonic band gap (PBG), based on the Transfer Matrix Method (TMM) and the two-fluid model. We show that the generalized Thue-Morse (GTM) and generalized Fibonacci GF(m, n) distributions provide a stacking of similar output multichannel with zero transmission when the input was a sharp resonance of peaks at given $n = 2pm$ where p, is a positive integer. Also, we consider GTM configuration and we apply a deformation $y = x^{h+1}$ along the PQC filter, which enhanced the band width of each channel with respect to the number of peaks inside the main transmittance. Here, the coefficient h represents the deformation degree, x and y are thicknesses of the layers before and after the deformation, respectively. This improves the characteristics of PBG.

Keywords: hybrid quasiperiodic PC, superconducting materials, GTM sequence, GF sequence, multichannel optical filters, deformed 1D photonic quasicrystals

1. Introduction

Photonic quasicrystals (PQCs) which are made of alternating dielectric and superconductor layers intervene in numerous researches due to their interesting optical properties [1–5]. This type of crystal is an artificial super lattice which is built according to quasiperiodic sequences. It is considerably different than photonic crystals (PCs) since it is a non-periodic structure with perfect long-range order and lack translational and it can be considered as an intermediate class between the random and periodic media. Our considered PQC consists of a stack of two different layers A and B which represent building blocks having a self-similarity distribution and long range order with no translational symmetry.

We mention that there are numerous examples of aperiodic chains constructed by a substitution rule. These chains allow forming many deterministic PQCs structures such as: Fibonacci, Thue-Morse, Rudin-Shapiro, Cantor, and Doubly periodic sequences.

Based on PQC heterostructure, many studies have been performed to carry out new optical devices. In this direction, the introduction of superconducting materials into the regular PQC photonic structure has been investigated in [5–7] in order to improve the characteristics of photonic band gap structures (PBGs) by changing the operating temperature of superconducting layers.

IntechOpen

Recently, 1D deterministic multilayered structure including superconducting layers have attracted much attention in developing new kinds of optical filters which make new PQCs devices for optoelectronic system [5, 8–11]. These quasiperiodic filters have been extended to thermally photonic crystals, including certain cascades superconducting/dielectric layers. It may be used in specific operations as specifying thermal sensors for remote sensing applications. In [12], the authors used superconductors instead of metals within the PC because of the damping of electromagnetic waves in metals. Moreover, the properties of PC including superconductors are mainly depending on the temperature T. In this chapter, based on hybrid dielectric/superconductor photonic quasicrystals, we develop a multichannel optical filter with tenability around two telecom wavelengths. The main multilayered stacks are organized following quasiperiodic sequences. Hence, a multitude of channel frequencies with zero transmission can be created inside the main photonic band gap (PBG), which offers a resonance state due to the specific defects insert along the structures.

The characteristics of PBGs depend on the parameters of sequences, the thickness of the superconductor and the operating temperature. Furthermore, a multitude of transmission peaks were created within the main PBG which shifted with temperature of superconductors and lattice parameters of the aperiodic sequence.

We also show that, by monitoring the parameters of GTM, the transmission spectrum exhibit at limited gaps a cutoff frequency which is sensitive to the temperature of superconducting layers. The properties of stop channel frequencies can be notably enhanced by applying a whole deformation $y = x^{h+1}$. Here, x is the main PQC and y the structure after deformation. It is found that the gaps broad in with the increase of h. Thus, the main structure can be used to design a useful tunable multichannel filter in the optical information field.

2. Problem formulation

In all this work, the photometric response (transmission and reflection) through the 1D photonic quasicrystal which contains superconductors, are determined by using the Transfer Matrix Method (TMM). We use also the theoretical Gorter-Casimir two-fluid model [13, 14] to describe the properties of the superconductor material ($YBa_2Cu_3O_7$).

The application of the two-fluid models and Maxwell's equations through, imply that the superconducting materials' electric field equation, obeys to the following equation:

$$\nabla^2 \mathrm{E} + {\mathrm{k_s}}^2 \mathrm{E} = 0 \qquad (1)$$

Where the wave number satisfies the corresponding equality:

$$\mathrm{k_s} = \sqrt{\frac{\omega^2}{\mathrm{c}^2} - \frac{1}{{\lambda_\mathrm{L}}^2}} \qquad (2)$$

with μ_0 and $c = 1/\sqrt{\mu_0 \varepsilon_0}$ denote the permeability and the speed of light in free space, respectively.

As mentioned above, the electromagnetic response of superconducting materials with the absence of an external magnetic field was defined by the Gorter-Casimir two-fluid models (GCTFM) in [13, 14]. According to GCTFM, the complex conductivity of a superconductor satisfies the following expression:

$$\sigma(\omega) = \frac{-ie^2 n_s}{m\omega} \tag{3}$$

Where n_s is the electron density and ω is the frequency of incident electromagnetic wave. Moreover, e and m represent the charge and the mass of electron, respectively. Under the approximation condition indicated in [14], the imaginary part of conductivity is given as follows:

$$\sigma(w) \approx \frac{-i}{\omega\mu_0\lambda_L^2(T)} \tag{4}$$

where λ_L signifies the term of London penetration depths and satisfies the following equality:

$$\lambda_L^2 = \frac{m}{\mu_0 n_s e^2}. \tag{5}$$

The complex conductivity is given by this formula: $\sigma = \sigma_1 - j\sigma_2$, where σ_1 and σ_2 are the real and imaginary parts of σ. Thus, the complex conductivity satisfies [14]:

$$\sigma_2 = \frac{1}{\omega\mu_0\lambda_L^2(T)}, \tag{6}$$

where ω is the operating frequency. The London temperature-dependent penetration depth is:

$$\lambda_L(T) = \frac{\lambda(0)}{\sqrt{1 - G(T)}} \tag{7}$$

Where $\lambda(0)$ denotes the London temperature penetration depth at T = 0 K, and G(T) is the Gorter-Casimir function. In this case, $G(T) = (T/T_c)^2$, where T_c and T are the critical and the operating temperatures of the superconductor, respectively.

Based on the Gorter-Casimir theory, we obtain that the relative permittivity of lossless superconductors takes the following equality [14]:

$$\varepsilon_s = 1 - \frac{\omega_{th}^2}{\omega^2}, \tag{8}$$

where ω_{th} is the threshold frequency of the bulk superconductor which satisfies: $\omega_{th}^2 =\sim c^2/\lambda^2$.

Then, the refractive index of the superconductor is written as follows:

$$n_s = \sqrt{\varepsilon_s} = \sqrt{1 - \frac{1}{\omega^2\mu_0\varepsilon_0\lambda_L^2}}, \tag{9}$$

In the following, the photometric response through the 1D photonic quasicrystal which contains superconductors, is extracted using the Transfer Matrix Method (TMM). This approach shows that the determination of the reflectance R and the transmittance T depends on refractive indices n_s and lower refractive indices n_d .

According to TMM, the transfer matrix C_j verifies the following expression [15]:

$$\begin{bmatrix} E_0^+ \\ E_0^- \end{bmatrix} = \prod_{i=1}^{m} \frac{C_j}{t_j} \begin{bmatrix} E_{m+1}^+ \\ E_{m+1}^- \end{bmatrix}, \tag{10}$$

For both TM and TE modes, C_j satisfies:

$$C_j = \begin{pmatrix} \exp\left(i\varphi_{j-1}\right) & r_j \exp\left(-i\varphi_{j-1}\right) \\ r_j \exp\left(i\varphi_{j-1}\right) & \exp\left(-i\varphi_{j-1}\right) \end{pmatrix}, \quad (11)$$

Where φ_{j-1} denotes the phase between the two succeed interfaces and it is given by the following formula

$$\varphi_{j-1} = \frac{2\pi}{\lambda} \hat{n}_{j-1} d_{j-1} \cos\theta_{j-1} \quad (12)$$

For the two polarizations (p) and (s), the Fresnel coefficients t_j and r_j take the following equalities [15]:

$$\begin{aligned} r_{jp} &= \frac{\hat{n}_{j-1}\cos\theta_j - \hat{n}_j\cos\theta_{j-1}}{\hat{n}_{j-1}\cos\theta_j + \hat{n}_j\cos\theta_{j-1}}; \quad t_{jp} = \frac{2\hat{n}_{j-1}\cos\theta_{j-1}}{\hat{n}_{j-1}\cos\theta_j + \hat{n}_j\cos\theta_{j-1}} \\ r_{js} &= \frac{\hat{n}_{j-1}\cos\theta_{j-1} - \hat{n}_j\cos\theta_j}{\hat{n}_{j-1}\cos\theta_{j-1} + \hat{n}_j\cos\theta_j}; \quad t_{js} = \frac{2\hat{n}_{j-1}\cos\theta_{j-1}}{\hat{n}_{j-1}\cos\theta_{j-1} + \hat{n}_j\cos\theta_j}, \end{aligned} \quad (13)$$

where $\hat{n}_j$ and θ_j are respectively the refractive indices and the angle of incidence in the jth layer which obey to the Snell's law: $\hat{n}_{j-1}\sin_{j-1} = \hat{n}_j\sin_j$ with $j\in[1, m+1]$.

Consequently, the transmittance satisfies [15]:

$$T_{rs} = \mathrm{Re}\left(\frac{\hat{n}_{m+1}\cos\theta_{m+1}}{\hat{n}_0\cos\theta_0}\right)|t_S|^2; \; T_{rp} = \mathrm{Re}\left(\frac{\hat{n}_{m+1}\cos\theta_{m+1}}{\hat{n}_0\cos\theta_0}\right)|t_P|^2, \quad (14)$$

3. Generalized quasiperiodic sequences

3.1 Generalized Thue-Morse sequence

A one dimensional GTM sequence is called aperiodic because it is more disordered than the quasiperiodic one. In addition, the two different materials included in one dimensional GTM system should be structured by applying the substitution rule: $\sigma_{GTM}(H, L) : H \rightarrow H^m L^n, L \rightarrow L^m H^n$ [16], where H and L represent the two layers, having the higher and the lower refractive indices, respectively. We note that the Fourier spectra of the GTM sequence is singular and continuous. Also, the GTM quasiperiodic chain is generated by a recursive deterministic sequence S_{k+1} verifying: $S_{k+1} = S_k^m \overline{S}_k^n$, where $\overline{S}_k^n$ is the conjugated sequence of S_k^n, m and n are the parameters of GTM sequence with order k. This rule can be applied to two dimensions: horizontally and vertically.

Based on GTM sequence S_{k+1}, we give **Table 1** which illustrates an example of organized multilayered stacks (H, L) for m = n = 2.

The configuration of the proposed 1D photonic dielectric/quasiperiodic superconducting layers which is built according to the GTM sequence is shown in **Figure 1**.

3.2 Generalized Fibonacci sequence

1D Fibonacci quasiperiodic sequences are constructed by applying the inflation rule in [17]: $\sigma_{GF}(H, L) : H \rightarrow HL, L \rightarrow H$ for the two blocks H and L, where H

Order of GTM	Organized $\{H, L\}$ blocks according GTM(2, 2) sequence
1	HHLL, with S_0 = H
2	HHLLHHLLLLHHLLHH
3	HHLLHHLLLLHHLLHHHHLLHHLLLLHHLLHH LLHHLLHHHHLLHHLLLLHHLLHHHHLLHHLL

Table 1.
Repeated {H,L} blocks determined by applying the substitution rule $\sigma_{GTM}(H, L)$.

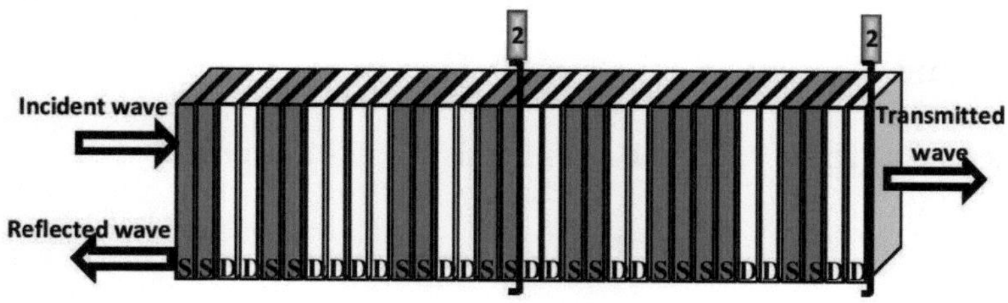

Figure 1.
Schematic drawing of 1D multilayered stacks made of dielectric (D)/superconducting materials (S), built according to the GTM(2, 2) sequence.

denotes the material with the higher refractive index, and L denotes the material with the lower refractive index. The GF chain is generated using the substitution rule: $\sigma_{GF}(H, L) : H \rightarrow H^m L^n, L \rightarrow H$. Thus, the GF sequence S_{k+1} satisfies the recursion relation: $S_{k+1} = S_k^m S_{k-1}^n$ with k is the order of GF sequence.

The Fourier transform of Fibonacci class of quasicrystal gives discrete values which represent the significant property of crystals. We note that the eigenvalues of related matrix Fibonacci spectrum are Pisot numbers. For the Fibonacci-type, the material waves interfere constructively in appropriate length. The analysis of Fibonacci quasicrystals submitted to X-ray diffraction shows a multitude of Bragg peaks. Moreover, quasicrystals which are based on the Fibonacci distribution ordered at long distances, show a typical construction without a forbidden symmetry. Hence, the generalized Fibonacci (GF) type gives some basic proprieties which are identical to those given by simple Fibonacci class such as Fourier spectrum with Bragg peaks, inflation symmetry and localized critical modes with zero transmission called pseudo band gaps. In a generic form of the organized multilayers (H, L) through Fibonacci sequence, the four multilayered stacks are grouped in **Table 2**.

As an example, the third order of GF(m, n) quasiperiodic photonic structure containing alternate dielectric (D) and superconducting layers (S) with m = n = 2 is shown in **Figure 2**.

Order of GF	Organized $\{H, L\}$ chain according GF(2, 2) sequence
1	HHLL, with $S_0 = L$ and $S_1 = H$
2	HHLLHHLLHH
3	H^2L^2 $H^2L^2HHH^2L^2H^2L^2HHH^2L^2H^2L^2$
4	$H^2L^2H^2L^2HHH^2L^2H^2L^2HHH^2L^2H^2L^2H^2L^2$ $H^2L^2HHH^2L^2H^2L^2HHH^2L^2H^2L^2H^2L^2H^2L^2HH$ $H^2L^2H^2L^2HH$

Table 2.
Generation of Fibonacci sequence and organized blocks (H, L) repeated by the substitution rule $\sigma_{GF}(H, L)$.

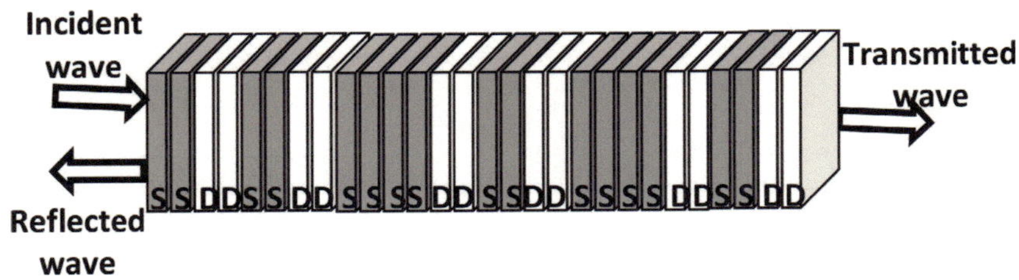

Figure 2.
Schematic representation showing the third generation of 1D GF(m, n) quasi-periodic multilayered stacks consisting of alternate dielectric (D)/superconducting materials (S).

4. Results and discussion

4.1 Multichannel filter narrow bands by using GTM sequence

4.1.1 Effect of GTM(m, n) parameters

In this subsection, we give the transmission properties of GTM and GF quasi-periodic one-dimensional photonic crystals (1DPCs) which contain

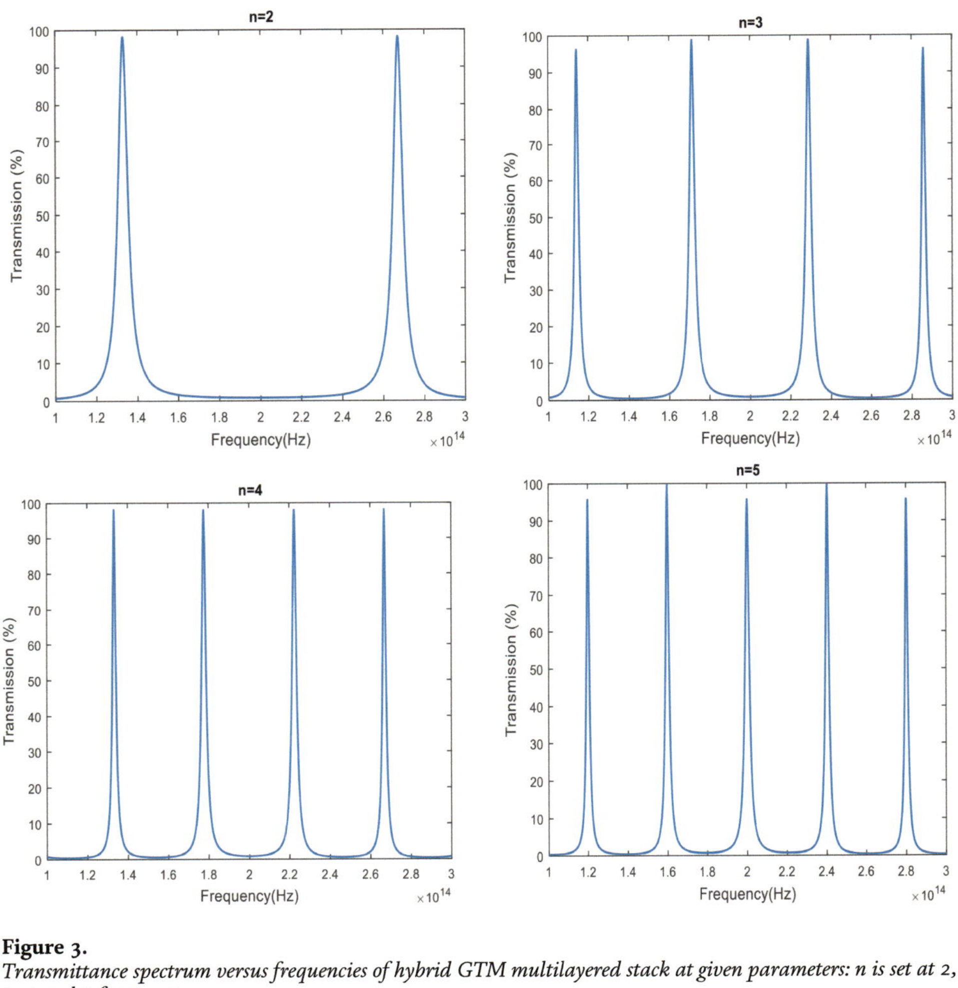

Figure 3.
Transmittance spectrum versus frequencies of hybrid GTM multilayered stack at given parameters: n is set at 2, 3, 4 and 5 for m = 2.

superconductors. We recall that our one dimensional photonic quasicrystal is made of alternating superconductors and dielectrics (SiO_2) with n_L = 1.45. In particular, the superconductor is assumed to be $YBa_2Cu_3O_7$ with a critical high-Tc temperature (Tc = 93 K) and a London penetration depth at zero temperature $\lambda_L(0) = \lambda_0 = 145\,\text{nm}$.

We adopt TMM approach to exhibit the transmittance, band gaps and characteristics of the hybrid GTM and GF photonic quasicrystals.

Figure 3 shows the transmittance spectrum, at normal incident angle for different n values.

We remark that the spectrum give a stacking of similar channels with zero transmission covering the whole frequency range. We also observe that the number of gaps increases with an increase of the lattice parameter n of GTM.

Also, sharp peaks of transmission appear for specific multiple frequencies. All peaks prohibit the stop band gaps and form a fine zone of propagation wave. This zone constitutes a little region of transmissions with small half bandwidth $\Delta f = 1.2$ THz. Similarly, the size of the output channels becomes narrow as n increases. Then, a large PBG zone was created. Thus, we note that the characteristics of channel filters are sensitive to lattice parameters of GTM sequence which organized the layers H and L. The similarity of transmission spectrum is caused by the self-similarity of geometrical GTM structures.

4.1.2 Effect of the thickness of superconductor on GTM structure

In this part, the superconductor's thickness is changed by varying the permittivity of its refractive index. **Figure 4** shows that a large PBG augments with an

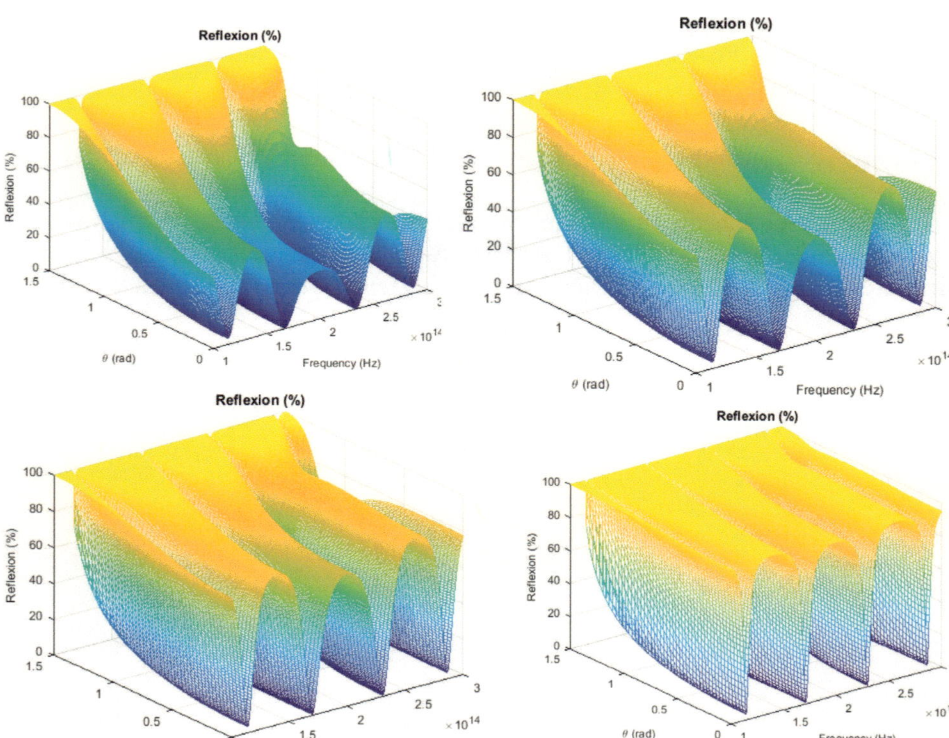

Figure 4.
The 3D reflectance spectrum through hybrid GTM(m, n) heterostructure at given values of superconductor's thicknesses: d_s is set at 20, 40, 60 and 80 nm.

augmentation the thickness. Full gaps were obtained for d_s = 80 nm. The amplitude of oscillations around the channels with T = 0 decreases with an increase of d_S. Also, a set of peaks is obtained for high values of thickness. Accordingly, the dip of each gap increases when the thickness of $YBa_2Cu_3O_7$ increases, and the pseudo PBG becomes a gap with zero transmission. This improves the characteristics of channel filters.

4.1.3 Quality factor (Q)

In this part, we calculate the quality factor based on the following formula: $Q = \sim f_c/\Delta f$, where Δf is the Full Width at Half Maximum (FWHM) of transmission peak and f_c is the wavelength of maximum transmission.

Our calculation is summarized in **Figure 5** which gives the evolution of quality factor Q versus the frequency center of resonant transmission peak for different superconductor temperatures T. We remark that Q is very sensitive to the position of resonant peaks in 170–171 THz frequency range and it is inversely proportional to superconductor's temperature T. The FWHM are approximately equal for the lower frequencies and it sharply increase for the higher frequencies range. Then, a high pass filter can be obtained for lower T.

In order to show the consequences of the variation of parameter p of GTM sequence, we determine the transmittance T versus the frequency for p = 7.

As it can be seen from **Figure 6**, the number of defect modes or channels depends on the superconductor's thicknesses and the distribution of layers. Moreover, the transmission spectrum exhibit a stacking of narrow gaps without oscillatory behavior. The bandwidth of each gap decreases regularly for an increase of parameter n and it probably forms a great wide PBG covering all telecommunication frequency range. The number of the transmission peaks increases as p increases. The band gaps are symmetrical about the separated transmission due to the symmetry of layers within the GTM structure.

4.1.4 Effect of superconductor temperature on GTM structure

In this subsection, we study the influence of superconductor's temperature on transmission spectrum of 1D hybrid GTM structure for different incidence levels. Thus, we evaluate the characteristics of multichannel. Indeed,

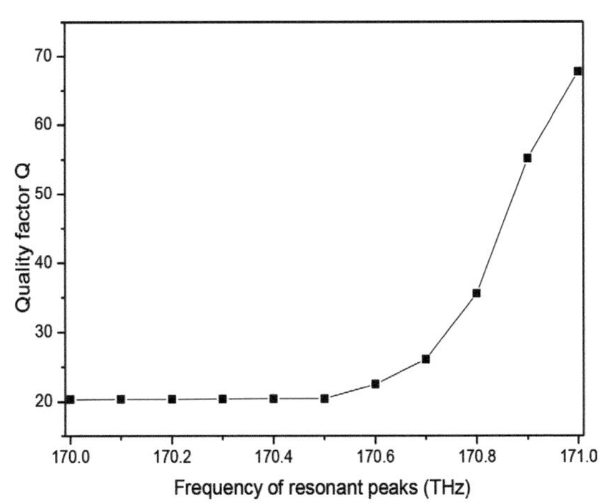

Figure 5.
Variation of factor quality Q of the GTM quasiperiodic multilayered stack containing a superconducting material versus frequency f (THz) at the frequency range between 170 and 171 THz.

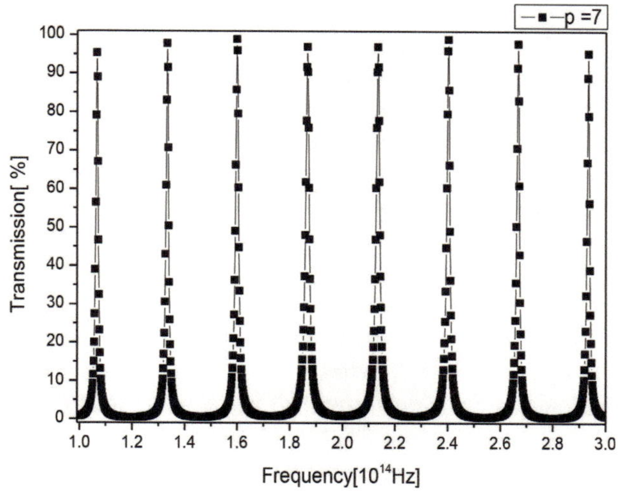

Figure 6.
A schematic view of transmittance spectra through the one dimensional photonic quasicrystals arranged according to GTM sequence for n = 2 and p = 7.

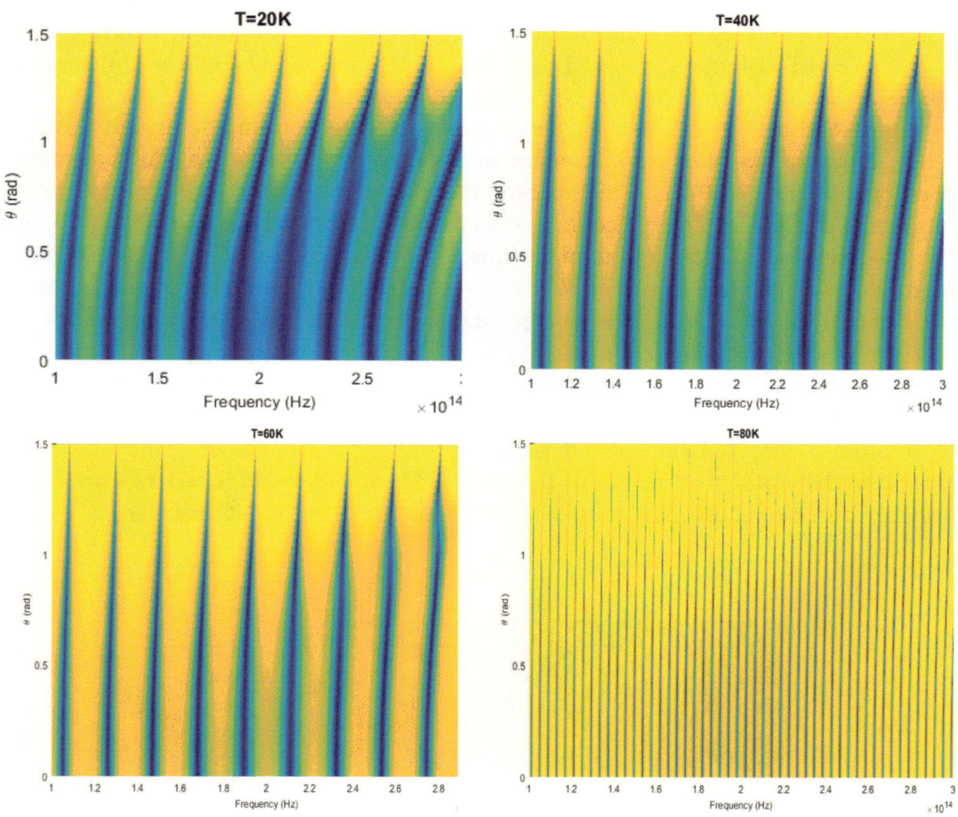

Figure 7.
A schematic view of distributed transmission of hybrid GTM photonic heterostructure versus *frequency and incident angle for T = 20, 40, 60 and 80 K.*

Figure 7 shows that GTM multilayer stack exhibits a specific zone with zero transmission (the yellow area) for different incident angles. In the corresponding band, the propagation wave is prohibited and reached the maximum recovers for θ = 1.5 rad.

Moreover, the spectrum presented a stack of band gaps and separated by sharp transmission peaks (the blue areas) allows the propagation of wave in this specific region of frequencies. The size of propagate zone within all PBG is sensitive to temperature T of $YBa_2Cu_3O_7$. The width of transmission peak within the channels increases progressively with the increase of T. A large zero of reflection bands is also noticed for T = 80 K, it covers all optical telecommunication frequency range and it constitutes perfect reflectors in these region.

4.1.5 Enhancement of PBGs by applying a particular deformation

In order to improve the characteristics of filtering channels, we apply a particular deformation h satisfying the following low $y = x^{h+1}$, where, x and y represent the coordinates of the main and the deformed GTM heterostructures, respectively.

We recall that in the main structure, two forms of layer, H and L are organized in a GTM sequence, where H and L are the superconductor and dielectric materials, respectively.

Then, the optical phase becomes: $\varphi_{j-1} = 2\pi/\lambda x'_0 \cos\theta_{j-1}$. Here, the optical thickness after deformation noted x'_0 satisfied the following form: $x'_0 = \lambda_0/4\left(j^{th} - (j-1)^{th}\right)$. In this case, j and λ_0 indicate the optical thickness of jth layer which depends on deformation value h and the reference wavelength. According to this notion, **Figure 8** illustrates the distributed of H and L with low and high refractive indices of the main and deformed multilayered stack. We take h = 0.1 and m = n = 2.

Figure 9 shows the reflectance spectrum for a corresponding deformed GTM heterostructure. For the optimum value of deformation, similar peaks of transmission appear inside all PBGs. This selective channel of transmission is sensitive to parameter n of GTM. The reflection bands form a typical output multichannel. Also, the number of channels and transmission peaks within PBGs increase when n augments. The channel of each PBG becomes narrow as n increases. In this case, m was maintained fix at 2. As a result, the characteristics of PBG are improved by applying the deformation h. Consequently, it is possible to improve the filtering properties by varying the suitable configuration of GTM parameters and the deformation h.

In order to improve the characteristics of filtering channels, we apply a deformation to the whole thicknesses of the main GTM structure. **Figure 10** shows the distributed of transmission versus frequency for varying deformation h.

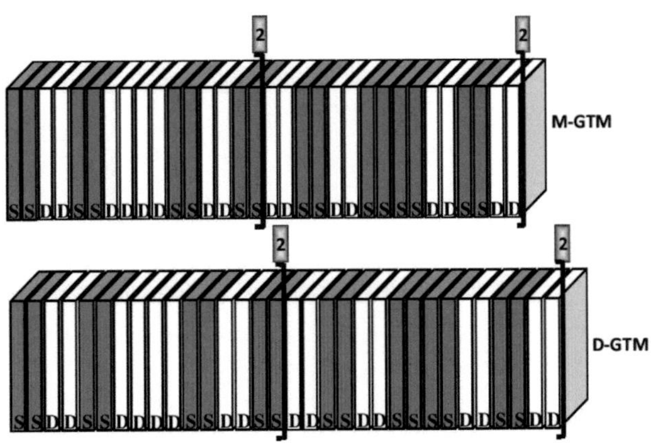

Figure 8.
Schematic representation of the main and deformed GTM photonic quasicrystals containing S and D materials, respectively. The deformation obeyed to the power law y = x^{h+1}.

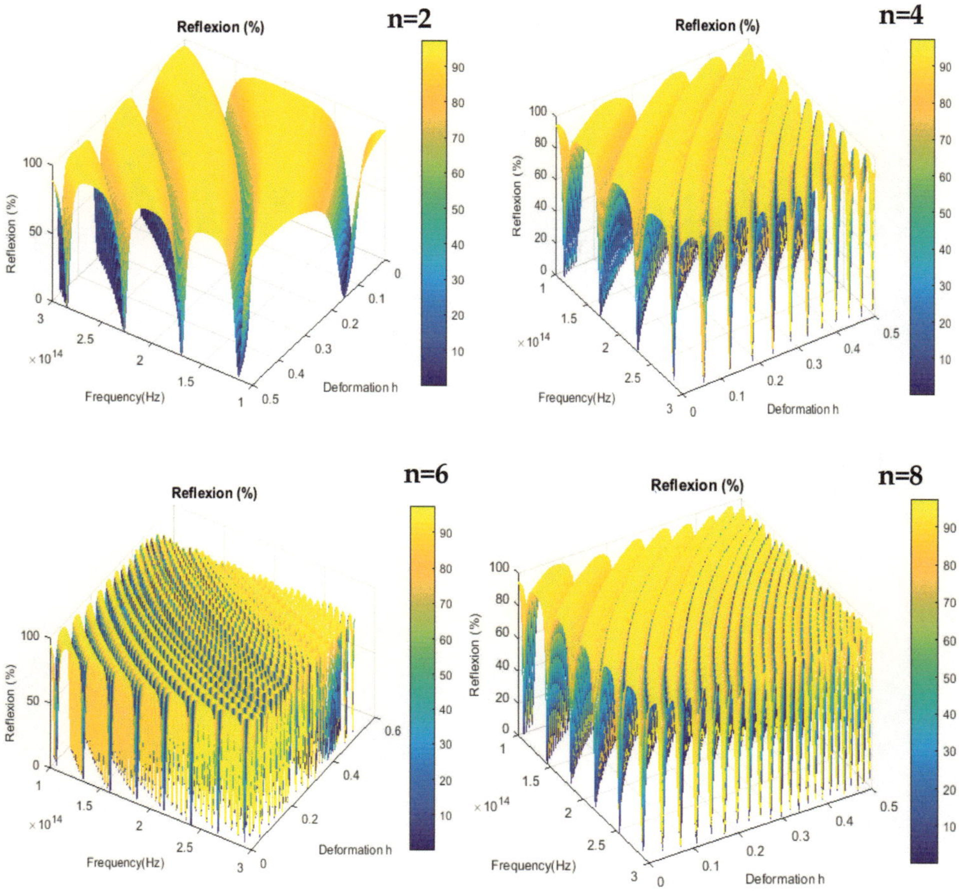

Figure 9.
3D reflectance spectrum at normal incidence from 1D GTM multilayered stack as a function of frequency (Hz) and deformation h with parameter n set to be 2, 4, 6, and 8.

The distribution of electric fields exhibits a stacking of a bending zone with zero transmission (yellow areas) and it is limited by harmonic peaks in blue fine zone. Then, these bending reflection bands are sensitive to superconductor's temperature. Thus, the PBGs are enhanced for an increase of T. In addition, the zone of transmission increases and the split peaks become narrow when T augments. Thus, the contrast indices of the two materials increases with T. Consequently, the considered factors cause broadening channels. Similarly, the intensity of transmission in all structures is reduced.

4.2 Generalized Fibonacci (GF) multichannel filters

4.2.1 The effect of GF(m, n) parameters

In this subsection, we study the properties of filtering through the 1D quasiperiodic GF multilayered stacks which contain superconducting materials. The considered common sequence suggests a typical aperiodic distribution of two alternating layers H and L with high and lower refractive indices, respectively.

The two constituent materials are arranged following the GF(m, n) sequence for m = pn, where p is a positive integer. We found that the transmission spectrum give similar band gaps which depend on the distributed layers initially fixed by the GF parameters (**Figure 11**). Therefore, the channel with zero transmission becomes

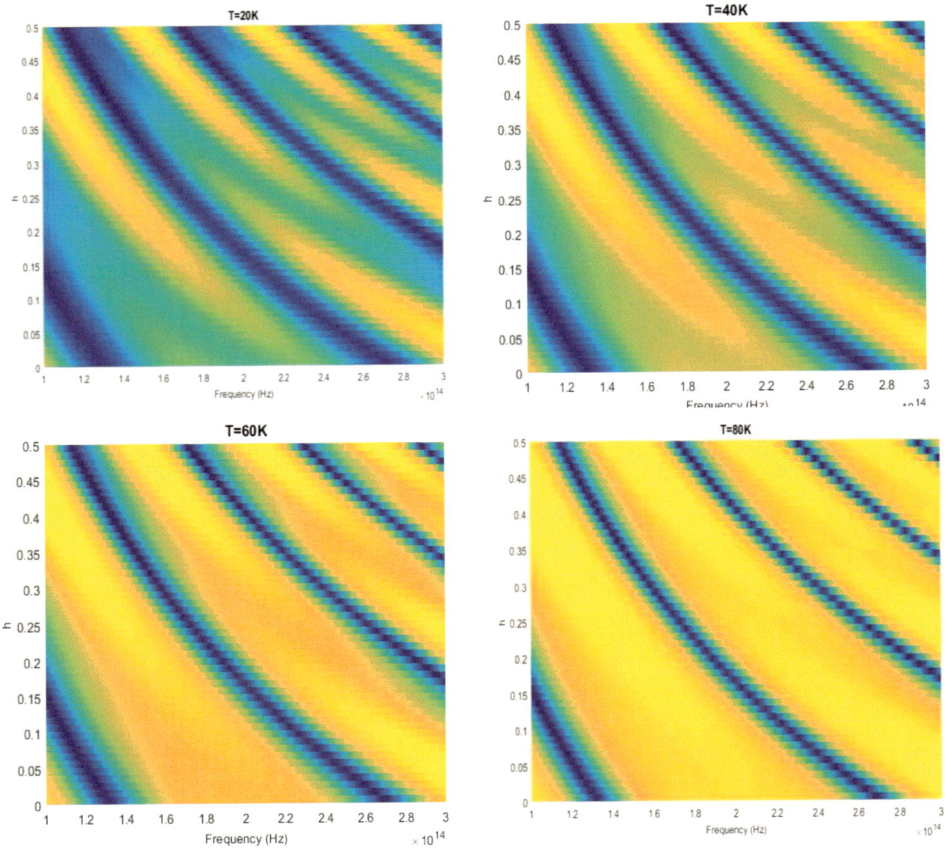

Figure 10.
A schematic view of distributed transmission of hybrid GTM multilayer stacks as a function of deformation degree h and frequency for different temperature values.

narrow when p increases. The hybrid GF heterostructure possess an oscillation transmission around all PBGs. Moreover, the stacking channels are symmetric around the reference frequency.

4.2.2 The effect of contrast indices on hybrid GF(m, n) system

In this subsection, we show the effect of the contrast indices between two alternating materials on the filtering properties. The contrast indices satisfy the following relation: $\Delta n = n_s - n_d$ with n_s and n_d represent the refractive indices of superconductor and dielectric, respectively. The same conditions are conserved to extract the transmission through the considered GF heterostructure.

Figure 12 gives the transmittance spectrum for different values of contract indices. We mention that the GF form exhibits a large frequency range with zero transmission and shows at limited gap a sharp transition from 0 to 1 at given Δn.

The intermediate point between inhibited and propagated waves indicates the cut-off frequency that allows the signal to propagate again, showing itself as a stop band filer. Moreover, we remark that the positions of the two cut-off frequencies f_{cL} and f_{cH} are very sensitive to the contrast indices. As long as Δn augments, the PBG increases similarly with the high cut-off frequency. Such interesting property may be applied to design a perfect reflector for high refractive index of superconductors. Thus, this type of reflectors exhibits a large bandwidth that contains the optical telecommunication frequency range.

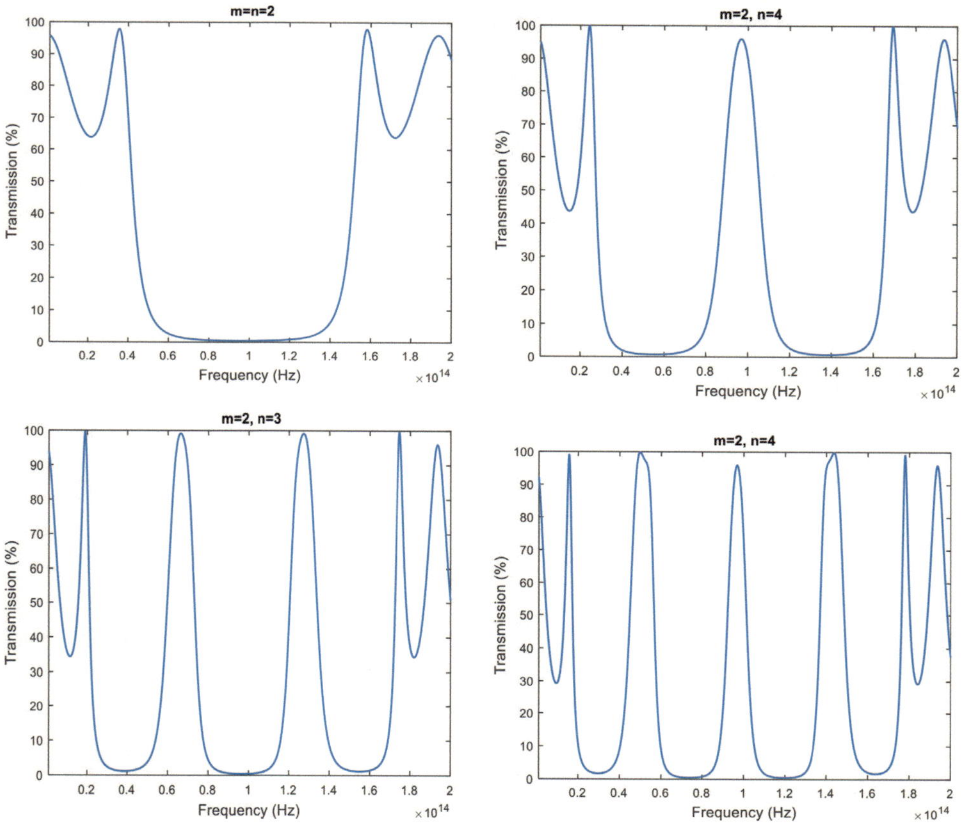

Figure 11.
Transmittance spectrums from 1D hybrid GF structure containing alternating dielectric/superconducting layers at given parameters: n set to be 2, 3 and 4 with m = 2.

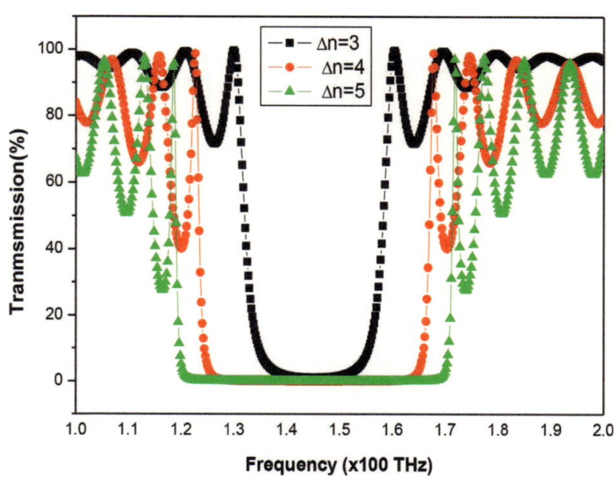

Figure 12.
Schematic representation of transmittance spectrums from 1D GF multilayered stacks at given Δn: n set to be 3,4 and 5 with n_d = 1.45.

5. Conclusion

The filtering properties of the 1D hybrid heterostructure built according to the GTM and GF sequences are investigated in this study. It was observed that the two common quasiperiodic sequences exhibits a multitude of channels with zero transmission for specific values of parameters m and n. In particular, the spectrum of GTM system possesses similar narrow gaps without oscillation beams at a given parameter: m = 2 pm. Indeed, a sharp transmission peak is appreciated in the whole frequency range whose positions are sensitive to superconductor temperature. Therefore, the considered system can be useful as a selective pass band multichannel filter whose narrow bandwidth can be adjusted by temperature. In addition, the main GTM system gives staking gaps which are enhanced by applying a specific deformation. Similarly, the GF heterostructure suggests an identical channel frequencies without transmission as compared to GTM system but their spectrum have particular oscillations around the cut-off frequency. Thus, the properties of filtering change by modifying the type of sequences and the parameters of constituent materials.

Author details

Youssef Trabelsi[1,2]

1 Photovoltaic and Semiconductor Materials Laboratory, El-Manar University-ENIT, Tunis, Tunisia

2 Department of Physics, King Khalid University, Abha, KSA

*Address all correspondence to: yousseff.trabelsi@gmail.com

References

[1] Zamani M. Spectral properties of all superconducting photonic crystals comprising pair of high-high, low-low or high-low temperature superconductors. Physica C: Superconductivity and Its Applications. 2016;**520**:42-46

[2] Srivastava SK. Study of defect modes in 1d photonic crystal structure containing high and low Tc superconductor as a defect layer. Journal of Superconductivity and Novel Magnetism. 2014;**27**:101-114

[3] Upadhyay M, Awasthi SK, Shiveshwari L, Srivastava PK, Ojha SP. Thermally tunable photonic filter for WDM networks using 1D superconductor dielectric photonic crystals. Journal of Superconductivity and Novel Magnetism. 2015;**28**: 2275-2280

[4] Rahimi H. Analysis of photonic spectra in Thue-Morse, double-period and Rudin-Shapiro quasiregular structures made of high temperature superconductors in visible range. Optical Materials. 2016;**57**:264-271

[5] Zhang HF, Liu SB, Yang H. Omnidirectional photonic band gap in one dimensional ternary superconductor-dielectric photonic crystals based on a new Thue Morse aperiodic structure. Journal of Superconductivity and Novel Magnetism. 2014;**27**:41-52

[6] Roshan Entezar S. Photonic crystal wedge as a tunable multichannel filter. Superlattices and Microstructures. 2015; **82**:33-39

[7] Lin WH, Wu CJ, Yang TJ, Chang SJ. Terahertz multichannel filter in a superconducting photonic crystal. Optics Express. 2010;**18**:27155-27166

[8] Zhang HF, Liu SB, Kong XK, Bian BR, Ma B. Enhancement of omnidirectional photonic bandgaps in one-dimensional superconductor dielectric photonic crystals with a staggered structure. Journal of Superconductivity and Novel Magnetism. 2013;**26**:77-85

[9] Barvestani J. Omnidirectional narrow bandpass filters based on one-dimensional superconductor dielectric photonic crystal heterostructures. Physica B. 2015;**457**:218-224

[10] Liu JW, Chang TW, Wu CJ. Filtering properties of photonic crystal dual-channel tunable filter containing superconducting defects. Journal of Superconductivity and Novel Magnetism. 2014;**27**:67-72

[11] Taherzadeh S, Vafafard A, Maleki MA, Mahmoudi M. Total reflection and transparent window in one-dimensional duplicated superconducting photonic crystal. Journal of Superconductivity and Novel Magnetism. 2013;**26**: 2911-2917

[12] Aly AH. Metallic and superconducting photonic crystal. Journal of Superconductivity and Novel Magnetism. 2008;**21**:421-425

[13] Tinkhman M. Introduction to Superconductivity. 2nd ed. New York: McGraw-Hill; 1996

[14] Mourachkine A. Room-Temperature Superconductivity. 1st ed. United Kingdom: Cambridge International Science Publishing; 2004

[15] Soltani O, Zaghdoudi J, Kanzari M. Analysis of transmittance properties in 1D hybrid dielectric photonic crystal containing, superconducting thin films. Physica B: Physics of Condensed Matter. 2018;**538**:62-69

[16] Moretti L, Rea I, Rotiroti L, Rendina I, Abbate G, Marino A, et al. Photonic band gaps analysis of Thue-Morse multilayers made of porous silicon. Optics Express. 2006;**14**:6264-6272

[17] Rostami A. Generalized Fibonacci quasiphotonic crystals and generation of superimposed Bragg Gratings for optical communication. Microelectronics Journal. 2006;**37**:897-903

Chapter 2

Synthesis of Curved Surface Plasmon Fields through Thin Metal Films in a Tandem Array

Gabriel Martinez Niconoff, Marco Antonio Torres Rodriguez, Mayra Vargas Morales and Patricia Martinez Vara

Abstract

We describe the generation of plasmonic modes that propagate in a curved trajectory inducing magnetic properties. This is performed by masking a metal surface with two screens containing a randomly distributed set of holes that follow a Gaussian statistic. The diameter of the holes is less than the wavelength of the illuminating plane wave. By implementing scaling and rotations on each screen, we control the correlation trajectory and generate long-range curved plasmonic modes. Using the evanescent character of the electric field, the study is implemented for the transmission of a plasmonic mode propagating in a tandem array of thin metal films offering the possibility to generate localization effects.

Keywords: plasmon mode, surface plasmon field, speckle, thin films, curved correlation trajectory

1. Introduction

During the last decade, the scientific community has shown an increasing interest in the models of plasmon fields due to their potential applications, which occur practically in all branches of science and technology. In the present study, we emphasize the analysis of correlation trajectories on a metal surface with random structure. The resulting model offers applications to development of nano-antennas having the possibility of a tunable bandwidth [1]. This type of structure has applications in the synthesis of new light sources and the control of magnetic effects [2]. The tunable effects are controlled with the curvature parameter having applications in surface-enhanced Raman spectroscopy (SERS), also as the local excitation of quantum dots. Implementing the evanescent behavior of the plasmon field, the analysis is extended to the propagation of plasmon fields through a tandem array of metal films similar to photonic crystal structures [3, 4].

As a starting point, we describe the study of the electric field in the neighborhood of a nanoparticle using the electrostatic approximation [2]. The electric field corresponds to the plasmon particle. This model allows the description of the interaction between two plasmon particles. The interaction is extended to describe the plasmon fields propagating on a surface generating a wave behavior satisfying the Helmholtz equation where the wave number must have complex values in order to recover the traditional surface plasmon models. Controlling the random distribution of nanoparticles, we analyze the correlation effects leading us to induce localization effects. This last statement is obtained by masking thin metal surface with two independent random array hole distributions. Controlling the scale factors, we modify the curvature of the correlation trajectory. The model is related with a speckle pattern emerging from a rough surface [5]. This configuration is similar to the configuration proposed by Reather for the coupling of plasmon fields. Experimental results are shown.

2. Analysis of plasmon particle

A nanoparticle is generated by a set of atoms; the plasmon particle corresponds with the surface current distribution of the atoms. The analysis is implemented applying the electrostatic approximation given by

$$\nabla^2 \phi = 0, \tag{1}$$

where ϕ is the potential function. Using variable separation in Cartesian coordinates on the $x - y$ plane, the equation acquires the form

$$\frac{\partial^2 \phi}{\partial x^2} + \frac{\partial^2 \phi}{\partial y^2} = 0. \tag{2}$$

Proposing the solution as

$$\phi = X(x)Y(y), \tag{3}$$

we obtain the equation system

$$\ddot{X} - \alpha^2 X = 0 \tag{4a}$$

$$\ddot{Y} + \alpha^2 Y = 0, \tag{4b}$$

where the coupling constant α is a complex number having the form $\alpha = a + ib$. This condition is necessary because perturbing the field, it must acquire a propagating behavior as it is shown below. Solving for X, we have

$$X = c_1 e^{cx} e^{idx} + c_2 e^{cx} e^{-idx}, \tag{5}$$

and the solution for Y is given by

$$Y = D_1 e^{icy} e^{-dy}. \tag{6}$$

Then, the complete solution ϕ acquires the form

$$\phi = A e^{cx} e^{idx} e^{-dy} e^{icy}, \tag{7}$$

with $c < 0$ and $d > 0$. Eq. (7) represents the boundary condition for the plasmonic field.

2.1 Description for the interaction between plasmon particles

The model is extended to describe the propagation of the electric field. For this, we propose that the electrostatic approximation is no longer fulfilled, acquiring the form of the Helmholtz equation having the form

$$\nabla^2\phi + k^2\phi = 0. \tag{8}$$

Looking for propagation along the x-coordinate, the equation acquires the form

$$\frac{\partial^2\phi}{\partial x^2} + \frac{\partial^2\phi}{\partial y^2} + k^2\phi = 0, \tag{9}$$

where k is the complex wave number $k = k_1 + ik_2$. Proposing a solution of the form $\phi = X(x)Y(y)$, we obtain the equation system given by

$$\ddot{X} + \left(k^2 - h^2\right)X = 0 \tag{10a}$$

$$Y + \alpha^2 Y = 0, \tag{10b}$$

whose solution acquires the form

$$\phi_p = Me^{\gamma x}e^{i\Omega x}e^{-dy}e^{icy}, \tag{11}$$

this equation must recover the structure of the electrostatic approximation for a single nanoparticle.

From the previous solution, it is easy to identify its behavior. Along the y-coordinate the field is bounded by the exponential term, which remains unperturbed by the presence of a second particle; the interaction occurs mainly in the x-coordinate. This behavior may be generalized acquiring a wave effect. A balance relation between the complex wave number k and the constant coupling α can be predicted; this interaction decreases the evanescent term, and the propagating term becomes dominant. This interaction is sketched in **Figure 1**.

In **Figure 1a**, the electrostatic approximation is valid for a single nanoparticle; the wave behavior is generated by another set of particles interacting shown in **Figure 1c**.

Until this point we have described the generation of a wave propagating in the x-coordinate; this analysis can be extended to the propagation in the $x - y$ plane, which is analyzed in the following section.

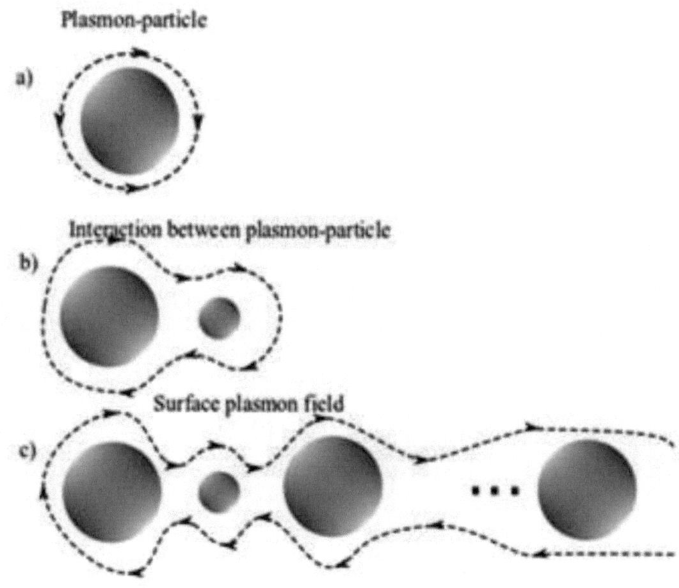

Figure 1.
(a) Localized electric field for a plasmon particle. (b) Interaction between two plasmon particles. (c) Sketch to describe the generation of a plasmon field in an array of nanoparticles.

3. Description statistics of correlation trajectories

In the present section, we describe the transfer of the statistical properties of an anisotropic two-dimensional random walk model to generate wave propagation on a metal surface, thus generating a curved surface plasmon mode. The model is conceptually simple. We describe a trajectory in a two-dimensional array, starting from a point P with coordinates $(0, 0)$. The random walk is characterized by a set of points randomly distributed, and the trajectory can be obtained from the correlation function corresponding to the flows of current probability. The statistical properties of a random distribution of points can be transferred to induce and control important physical effects. For example, it is known that the amplitude distribution of a speckle pattern follows Gaussian statistics [6, 7]. The statistic of the speckle pattern is matched with a random hole distribution, and it is transferred on a metal surface. The analysis is obtained by masking the surface metal which is considered to be formed by a set of square cells. The probability of a hole being present at the center of each cell is P; therefore, the probability of the absence of a hole is $(1 - P)$. The surface contains N cells, and the probability of the surface contains n-holes, assuming that a Bernoulli distribution is

$$P(n) = \binom{N}{n} P^n (1 - P)^{N-n}. \tag{12}$$

When the number of cells N increases, the Bernoulli distribution tends to a Gaussian distribution of the form

$$\rho(x, y) = \frac{1}{\sqrt{2\pi\sigma^2}} e^{-\frac{x^2+y^2}{2\sigma^2}}, \tag{13}$$

where σ^2 is the variance. Interesting features can be identified by describing the self-correlation in this type of distribution. The simplest case occurs when two screens are superposed and, subsequently, one of them is rotated by a small angle. In order to understand the generation of the self-correlation trajectory, we focus on a single hole. In this case, it is evident that the hole follows a circular arc by joining all the points of constant probability and the complete correlation trajectory is a circle. The result in this case is shown in **Figure 2a**. The correlation trajectory can be controlled by inducing a scale factor in the distribution of random points. By

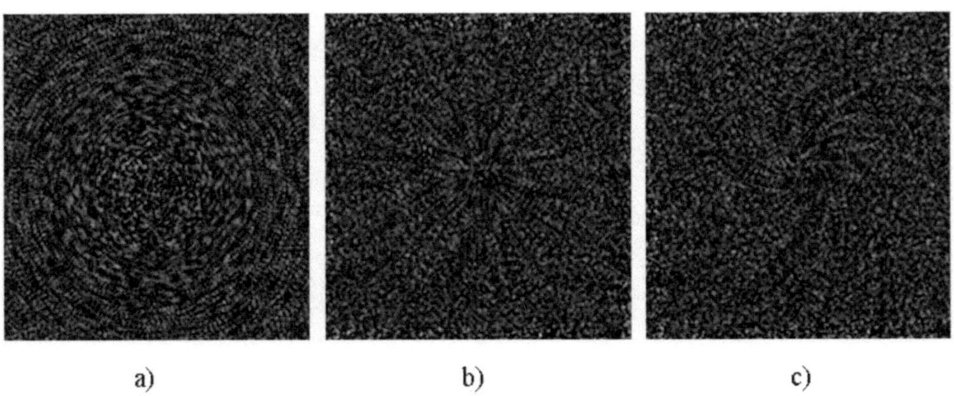

Figure 2.
(a) Set of points following a Gaussian distribution. (b) Correlation function between two Gaussian sets of points where one mask was rotated by a small angle. (c) Probability flow trajectories between two mask Gaussian points, one of them is scaled by approximately 95%, without rotation. (d) Same as in (c) but with a rotation of approximately 5°.

superposing the two screens again, it is evident that the scale factor shifts the point along a linear trajectory perpendicular to the regions of constant probability, which are sets of circles, as deduced from the argument of the Gaussian distribution. The analysis is presented in an equivalent way for a speckle pattern using the fact that both of them have the same probability distribution. In **Figure 2b**, we show these correlation trajectories. Finally, by introducing a small rotation, the linear trajectories are curved, as shown in **Figure 2c**.

This result can be explained as follows: the correlation function of two scaled and rotated surfaces have the form

$$\rho_1(x,y)\cdot\rho_2(x',y') = \frac{1}{\sqrt{2\pi\sigma_1\sigma_2}}\exp\left\{-\frac{x^2+y^2}{2\sigma_1^2}\right\} \times \exp\left\{-\left[\frac{d(xcos\theta+ysen\theta]^2+[d(-xsen\theta+ycos\theta)]^2}{2\sigma_2^2}\right\} \tag{14}$$

Analyzing the argument of the exponential function as a quadratic form, it can be shown that the curves of constant correlation are ellipses, presenting a reference system where they acquire the canonical form

$$\frac{x^2}{a^2}+\frac{y^2}{b^2}=1. \tag{15}$$

The probability flows through the orthogonal trajectories between the two regions of constant probability, whose differential equation is given by

$$y' = \frac{b^2 y}{a^2 x}. \tag{16}$$

Further, the corresponding solution is given by

$$y = cx^{\alpha}, \tag{17}$$

where c is an arbitrary constant and $\alpha = \frac{b^2}{a^2}$, which carries the information about the scale between the two probabilistic processes.

3.1 Graphical description and experimental implementation of the correlation trajectory

A fundamental part of the chapter consists of describing a method to generate surface plasmon fields propagating along predetermined trajectories. This can be obtained analyzing the correlation function between two screens where each one has a random hole distribution following a predetermined probability density function. This method has the characteristic that the correlation trajectory geometry presents a tunable curvature which allows the possibility to generate long-range surface plasmon.

An alternative model to generate the curved correlation trajectories is performed using a speckle pattern as it is shown in **Figure 4**.

The optical system that rotates the image can be a prism-type Dove. Modifying the illumination configuration using a convergent beam and changing the relative distance between the two speckle patterns obtained by shifting one mirror a scale factor are introduced. The irradiance superposition between the two speckle patterns generates the desired correlation trajectories. The speckle pattern is shown in **Figure 3**.

It is known that the irradiance function for the speckle pattern has associated a probability density function-type exponential decreasing function. The decreasing

Figure 3.
Speckle pattern generated with a rough surface illuminated with a plane wave.

term can be matched with the decaying ratio of the plasmon mode. This configuration allows improving the generation of plasmon field avoiding the masking of the metal surface which must be made with lithography techniques. These comments represent novel applications of the speckle pattern.

The correlation trajectories generated will be implemented in the following section to describe the surface plasmon. By the fact that the correlation occurs in a curved trajectory, we expect the surface plasmon to present a magnetic behavior.

3.2 Generation of curved surface plasmon modes

The previous statistical description will be employed for the synthesis of surface plasmonic modes. The expression for the electric field of an elementary surface plasmonic mode propagating along the z-axis is given by

$$E(x,z) = \left(\hat{i}a + \hat{k}b\right) \exp\{-\alpha x\} \exp\{i\beta z\}, \tag{18}$$

where $\beta = \frac{w}{c}\left(\frac{\varepsilon_1\varepsilon_2}{\varepsilon_1+\varepsilon_2}\right)^{1/2} = \xi + i\eta$ is the dispersion relation function and ε_1, ε_2 represent the permittivity of the dielectric and metal, respectively. Rotating the reference system along the x-axis, the elementary surface plasmon mode acquires the form

$$E(x,z) = \left(\hat{i}a + \hat{j}bsin\theta + \hat{k}bcos\theta\right) \times \exp\{-\alpha_1 x\} \exp\{i\beta(zcos\theta + ysin\theta)\}. \tag{19}$$

Using the functional relation given by Eq. (17), the expression for the curved plasmonic mode is given by

$$E(x,y) = \left(\hat{i}a + \hat{j}bsin\theta + \hat{k}bcos\theta\right) \times \exp\{-\alpha_1 x\} \exp\{i\beta(y^{\alpha}cos\theta + ysin\theta)\}. \tag{20}$$

By means of the Maxwell equations, we can obtain the expression for the magnetic field and the energy flux given by the Poynting vector.

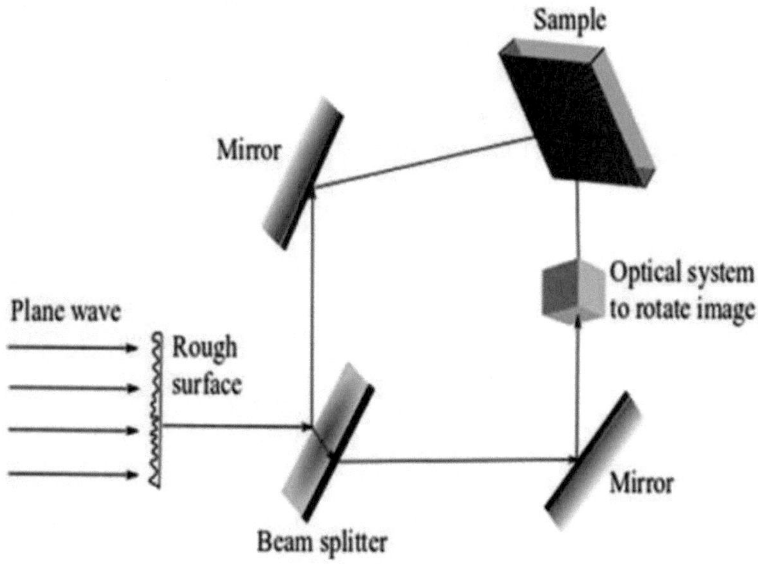

Figure 4.
Experimental setup to generate speckle correlation trajectories.

For the experimental setup, we propose to illuminate a thin flat Au film (thickness ~20–40 nm) with a correlated speckle pattern as shown in **Figure 4**. The illumination consists in two speckle patterns: each one is visualized as a set of circular motes randomly distributed following a Gaussian probability density function. The wavelength is $\lambda = 1550$ nm. The geometrical parameters are agreeing with those reported in [8]. The correlation curve corresponds to the surface plasmonic mode given by Eq. (20). Notably, the statistical properties of the speckle pattern are transferred to the metal surface as the plasmonic mode propagating along the correlation trajectory. In order to allow the generation of a long-range curved plasmonic mode, the correlation length must be less than 2μ to guarantee resonance effects [9, 10]; this can be controlled with the roughness parameters of the surface implemented to generate the speckle pattern avoiding the power decay along the correlation trajectory. The experimental setup is sketched in **Figure 5**.

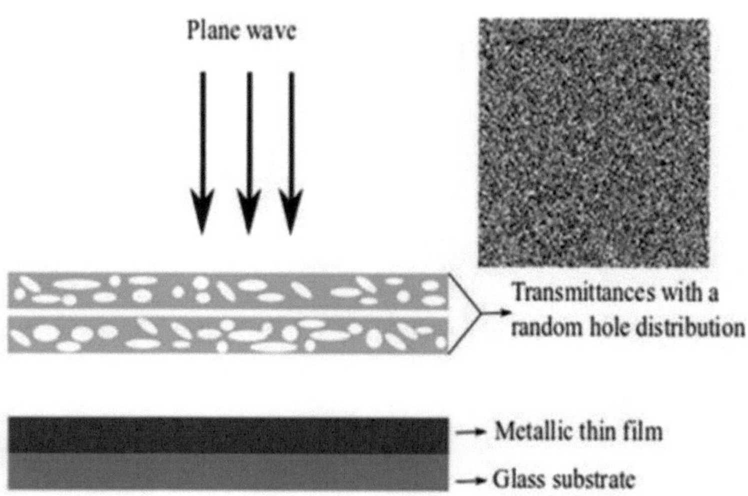

Figure 5.
Masked metal surface: The typical wavelength is IR.

The analysis presented can be extended to other plasmonic configurations which are presented in the following section.

4. Propagation in a tandem array of thin metal films

The natural extension of the analysis presented is the transfer of the plasmonic mode to a tandem array of thin metal surface, shown in **Figure 6**. This is possible using the evanescent behavior along the x-axis of the curved surface plasmon field. This behavior has been implemented to generate an optical field redistribution propagating along an optical waveguide array [11]. In this model, the evanescent character is used to tunnel the optical field.

The transmission of the plasmonic mode satisfies the following system of differential equations:

$$i\frac{dE_n}{dz} + \beta E_n + C_{n+1}E_{n+1} + C_{n-1}E_{n-1} = 0 \tag{21a}$$

$$n = 1, 2, 3\ldots, \tag{21b}$$

where β is the dispersion relation function and C_i represents the coupling constant, which depends on the relative separation between neighborhood surfaces [12]. The solution of the previous equation is similar to that presented in [11]; however, to associate a physical meaning to the coupling constant C_i, we present the analysis of two thin metal films.

The simplest case occurs when the system is formed by two thin metal films separated by a dielectric medium whose thickness must be less than 50 nm. The evanescent decay depends on the modulus of the permittivity quotient [13], and at this thickness is possible to generate tunneling effects [11]. Subsequently, the system of Eq. (21a) acquires the simple form

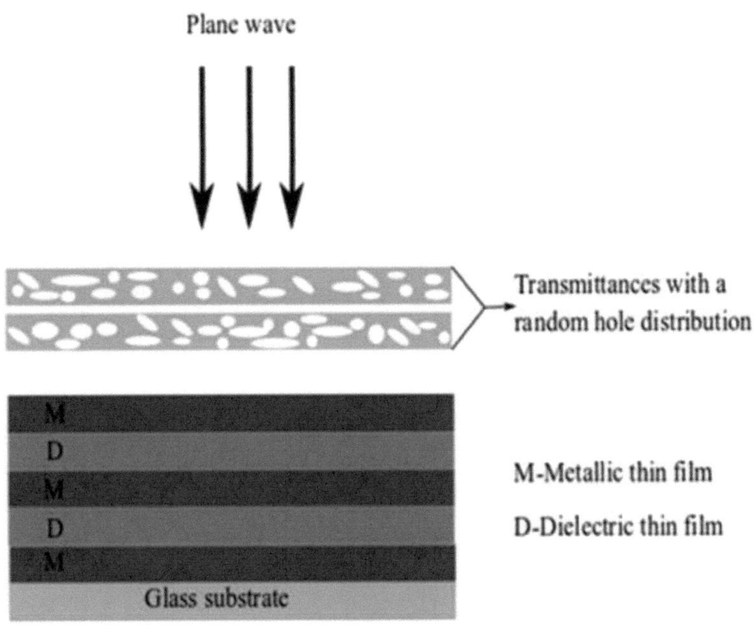

Figure 6.
Tandem array to propagate the plasmon field: the width of the metal is 20 − 40 *nm and that dielectric film is* 20 − 40 *nm.*

$$i\frac{dE_1}{dz}+\beta E_1+C_2E_2=0, \tag{22a}$$

$$i\frac{dE_2}{dz}+\beta E_2+C_1E_1=0. \tag{22b}$$

Rewriting it in matrix form, we obtain

$$i\begin{pmatrix} \frac{dE_1}{dz} \\ \frac{dE_2}{dz} \end{pmatrix}=-\begin{pmatrix} \beta & c_2 \\ c_1 & \beta \end{pmatrix}. \tag{23}$$

It can be deduced that, as a consequence of the energy conservation, the matrix structure must be symmetric. This indicates that $c_1=c_2=c$, and the general solution is

$$\begin{pmatrix} E_1 \\ E_2 \end{pmatrix}=d_1\begin{pmatrix} \xi_1 \\ \xi_2 \end{pmatrix}\exp\left(\lambda_1 z\right)+d_2\begin{pmatrix} \eta_1 \\ \eta_2 \end{pmatrix}\exp\left(\lambda_2 z\right), \tag{24}$$

where d_i represents arbitrary constants and $\xi_{1,2}$ and $\eta_{1,2}$ represent the eigenvectors with eigenvalues $\lambda_{1,2}$ satisfying the characteristic equation depending on the coupling constant:

$$\lambda_{1,2}=\beta\pm c. \tag{25}$$

Moreover, it is known that the eigenvectors must be complex [14]. Subsequently, without loss of generality, the solution can be rewritten as

$$\begin{pmatrix} E_1 \\ E_2 \end{pmatrix}=d_1\begin{pmatrix} 1 \\ i \end{pmatrix}\exp\left(\lambda_1 z\right), \tag{26}$$

which indicates that the shift generated between each plasmon mode presents similar features as the coupling mode theory [12]. This analysis leads to the expression for the plasmonic mode as

$$E_1=A\ \vec{\xi}\ \exp\left(-|\alpha x|\right)\exp\left(i\beta s\right) \tag{27a}$$

$$E_2=iA\ \vec{\xi}\ \exp\left(-|\alpha x|\right)\exp\left(i\beta s\right), \tag{27b}$$

where $\vec{\xi}$ is a unit vector tangent to the correlation curve and s is the arc length on the same curve; we remark that the correlation trajectory is given by Eq. (20).

Eq. (24) describes the evanescent coupling through a tandem array of thin metal films. Notably, the boundary conditions of the electric field indicate that the geometry of the plasmon field generated in the first thin metal film must be preserved in all the surfaces. This shows that the transmission of the curved plasmonic mode allows inducing magnetic properties in the system [15–18].

5. Conclusions

The statistical properties of the distribution of random holes or equivalently the speckle pattern were transferred to a metal surface to stablish the conditions to generate long-range curved plasmonic modes. In the case of hole distribution, this can be implemented by masking a thin metal film with two screens that allows controlling the correlation trajectory whose geometry corresponds to a curved long-range surface plasmonic mode. Another possibility was illuminating the metal thin film with two correlated speckle patterns. An important consequence of these configurations is that the set of curved surface plasmonic modes presents a vortex structure that allows to induce magnetic properties [17]. Using the evanescent character of the plasmon modes, the electric field was transferred to the propagation in a tandem array of thin metal films offering applications to design photonic crystals with tunable and localized magnetic properties.

The theoretical point of view presented in this study allows incorporating other effects such as percolation effects which consist in propagating the electric field through random structures. The main characteristic is that the plasmon field presents fractal properties which are the origin of interesting magnetic properties implicit in the curved trajectory of the set of plasmonic modes; more details can be found in [18]. The model presented can be extended by implementing different hole distribution geometries which modify the plasmonic resonance effects. Notably, the curved trajectories have associated focusing regions, and, subsequently, the corresponding magnetic singularity offers the possibility of implementation in the generation of plasmonic magnetic mirrors.

Finally, we remark that the analysis presented offers applications to photonic crystal as a metamaterial design [19–23] since breaking the periodicity or incorporating another type of metal on a selected region is similar to doping the structure and then is possible to induce localization effects. The excitation of plasmon fields using a speckle patterns offers the possibility to incorporate the tunable behavior of the correlation trajectory offering interesting applications in the development of plasmonic antennas and synthesis of accelerating plasmon modes [21], extending the plasmonic optical models.

Acknowledgements

The authors MATR and MVM are very grateful to CONACyT for their support.

Author details

Gabriel Martinez Niconoff[1*], Marco Antonio Torres Rodriguez[1], Mayra Vargas Morales[1] and Patricia Martinez Vara[2]

1 Departamento de Optica, Instituto Nacional de Astrofisica, Optica y Electronica (INAOE), Puebla, Mexico

2 Departamento de Ingenierias, Benemerita Universida Autonoma de Puebla (BUAP), Puebla, Mexico

*Address all correspondence to: gmartin@inaoep.mx

References

[1] Brinks D, Castro-Lopez M, Hildner R, Van Hulst NF. Plasmonic antennas as design elements for coherent ultrafast nanophotonics. Proceedings of the National Academy of Sciences. 2013; **110**(46):18386-18390

[2] Aroca R. Surface-Enhanced Vibrational Spectroscopy. J. Wiley and Sons; 2006

[3] Berger V. Nonlinear photonic crystals. Physical Review Letters. 1998; **81**:4136-4139

[4] O'Brien S, Pendry JB. Photonic band-gap effects and magnetic activity in dielectric composites. Journal of Physics: Condensed Matter. 2002;**14**(15):4035-4044

[5] Raether H. Surface Plasmons on Smooth and Rough Surfaces on Gratings. Berlin Heidelberg: Springer-Verlag; 1988

[6] Spitzer F. Principles of Random Walk. New York: Springer-Verlag; 2001

[7] Goodman JW. Statistical properties of laser speckle patterns. In: Laser Speckle and Related Phenomena. Berlin Heidelberg: Springer; 1975. pp. 9-75

[8] Martínez Niconoff G, Martínez Vara P, Diaz Gonzalez G, Silva Barranco J, Carbajal Dominguez A. Surface plasmon singularities. International Journal of Optics Special Issue of Nanoplasmonics and Metamaterials. 2012;**2012**:7

[9] Jung K-Y et al. Au/SiO_2 Nanoring Plasmon waveguides at optical communication band. Journal of Lightwave Technology. 2007;**25**(9): 2757-2765

[10] Ahmadivand1 A, Golmohammadi S. Electromagnetic plasmon propagation and coupling through gold nanoring heptamers: A route to design optimized telecommunication photonic nanostructures. Applied Optics. 2004; **53**(18):3832-3849

[11] Martin L, Di Giuseppe G, Perez Leija A, Keil R, Dreisow F, Heinrich M, et al. Anderson localization in optical waveguide arrays with off-diagonal coupling disorder. Optics Express. 2011; **19**(14):13636-13646

[12] Rodrıguez-Lara BM, Soto-Eguibar F, Zarate Cardenas A, Moya-Cessa HM. A classical simulation of nonlinear Jaynes–Cummings and Rabi models in photonic lattices. Optics Express. 2013;**21**(10): 12888

[13] Lee SY, Park J, Kang M, Lee B. Highly efficient plasmonic interconnector based on the asymmetric junction between metal-dielectric-metal and dielectric slab waveguides. Optics Express. 2011;**19**(10):9562-9574

[14] Hirsch MW, Smale S, Devaney RL. Differential Equations, Dynamical Systems, and an Introduction to Chaos. USA: Academic press, Elsevier; 2012

[15] McGurn AR, Maradudin AA, Celli V. Localization effects in the scattering of light from a randomly rough grating. Physical Review B. 1985;**31**(8):4866

[16] Zia R, Schuller JA, Brongersma ML. Near-field characterization of guided polariton propagation and cutoff in surface plasmon waveguides. Physics Review B. 2006;**74**:165415

[17] Engelhardt M, Langfeld K, Reinhardt H, Tennert O. Deconfinement in SU(2) Yang-Mills theory as a center vortex percolation transition. Physics Review D. 2000;**61**: 054504

[18] Enders D, Nagao T, Pucci A, Nakayama T, Aono M. Surface-enhanced ATR-IR spectroscopy with

interface-grown plasmonic gold-island films near the percolation threshold. Physical Chemistry Chemical Physics. 2011;**13**(11):4935-4941

[19] Maigyte L, Staliunas K. Spatial filtering with photonic crystals. Applied Physics Reviews. 2015;**2**(1):011102

[20] Zheng X, Smith W, Jackson J, Moran B, Cui H, Chen D, et al. Multiscale metallic metamaterials. Nature Materials. 2016;**15**:1100-1106

[21] Zhang P, Hu Y, Cannan D, Salandrino A, Li T, Morandotti R, et al. Generation of linear and nonlinear nonparaxial accelerating beams. Optics Letters. 2012;**37**(14):2820-2822

[22] Yeh P. Introduction to Photorefractive Nonlinear Optics. New York: J. Wiley and Sons; 1993. pp. 47-61

[23] Prasad P. Nanophotonics. Hoboken, New Jersey: J. Wiley and sons; 2004

Chapter 3

Localized Excitation of Single Atom to a Rydberg State with Structured Laser Beam for Quantum Information

Leila Mashhadi and Gholamreza Shayeganrad

Abstract

Sufficient control over the excitation of the Rydberg atom as a quantum memory is crucial for the fast and deterministic preparation and manipulation of the quantum information. Considering the Laguerre-Gaussian (LG) beam spatial features, localized excitation of a four-level atom to a highly excited Rydberg state is presented. The position-dependent AC-Stark shift of the first and Rydberg state in the effective quadrupole two-level description of a far-detuned three-photon Rydberg excitation results in a steep trapping potential for Rydberg state. The transfer of optical orbital angular momentum from LG beam to the Rydberg state via quadrupole transition in the last Rydberg excitation process offers a long-lived and controllable qudit quantum memory. The effective quadrupole Rabi frequency is presented as a function of ratio of the first to Rydberg excitation laser beam waist and the center of mass position inside the trap. It depicts high accuracy of detecting Rydberg atom at the center of the trap, which can pave the way for implementation of high-fidelity qudit gate.

Keywords: Rydberg excitation, dipole-quadrupole trap, Laguerre-Gaussian beam, qudit

1. Introduction

1.1 Trapped neutral atom as quantum memories

Photons are ideal carrier of quantum information for communication, but storing them for a long time is difficult. In optical quantum communication, a quantum network can consist of spatially separated quantum memories to store and manipulate information which is encoded in internal quantum state of physical system such as photons, ions, atoms, etc. [1]. In order to operate a successful long-distance quantum communication and quantum information processing, the quantum memory should have a high-storage efficiency, which is defined as the ratio of input photon energy to long coherence time. Different physical systems such as cold atoms trapped in optical lattices [2–8] or cold ions in electrostatic traps [9] are used as quantum memories. Ions are electrically charged and therefore can be tightly confined in deep traps for a long time. However, the strong Coulomb repulsion limits the number of ions that can be precisely controlled in a single trap. In contrast, more promising experiments could be made with the cold trapped neutral atoms. The neutral atoms usually interact only at very short range and can be collected in large ensembles without perturbing each other, and therefore the decoherence and losing the quantum information due to the interaction with the environment are low. In other words, single qubits [1] or even multi-qubits can be encoded in atomic states which afford long coherence times. Cooling and trapping of neutral atoms is one of the challenging techniques to achieve a higher signal-to-noise ratio and to stabilize the system over long periods for more balanced memory efficiency and fully control of all physical degrees of freedom with long coherence times. To keep the atom in trap, it is necessary to raise the atomic trap depth to be comparable to or even larger than the thermal energy of background atoms. Although, in principle, there is no fundamental limit to the lifetime with sufficiently deep traps, there are practical limitations. Optical traps for neutral atoms cannot be arbitrarily deep since both the trap depth and the photon scattering rate, which causes decoherence, scale proportional to the optical power. Even in the absence of collisional losses as the main reason to lose the trapped atom, the heating due to fluctuations of the trapping potential can eventually cause trapped atoms to escape. As a result, the trapping system used to confine the atoms is switched off during the storage. Nevertheless, the residual electric and magnetic field still limits the lifetime of the quantum memory [10, 11]. For a quantum memory in a magneto-optical trap (MOT), the atomic diffusion imposes a strict limitation on the storage time, which in turn limits the maximum distance for quantum communication in practical applications. On the contrary, the optical dipole trap can offer an array of spatial forms which can be rapidly switched as well [12]. Optical dipole traps rely on the induced atomic electric dipole interaction with electric component of the trapping light. The power needed for trapping depends on the desired trap depth and the detuning from the nearest optical transitions. Small detuning gives deeper traps, with a depth scaling as $1/\Delta$, where Δ is the detuning from the nearest strong electronic transition. This must be balanced against the photon scattering rate, which causes heating and decoherence and scales as $1/\Delta^2$ [13]. When the light frequency is far-detuned [12, 14, 15], a nearly conservative potential well with less influence from spontaneous photon scattering is created for the atom. The AC-Stark shift in the atomic state from the induced dipole gives sensitivity to intensity noise of lasers and atomic position. The nonuniform spatial intensity profile results in the intensity-dependent AC-Stark shift and defines the shape of associated atom trap. Therefore, the AC-Stark shift affects the qubit level, and fluctuation in laser field leads to broad the qubit line.

In red-detuned far-off resonance optical dipole trap (FORT) [16, 17], the laser beam is focused to attract atoms to the region, where the intensity is high. The large detuning efficiently suppresses the effect of scattered photons. On the other hand, the blue-detuned FORT [18, 19] can confine the atom in the dark region of the blue-detuned laser beams. The blue-detuned FORT has several advantages over the bright trap of the red-detuned laser. As the atoms are trapped in the "dark" place, the photon scattering rate due to the trapping laser can be greatly reduced, while in the bright trap, this rate can be reduced only by increasing the detuning of the trapping laser. A quantum memory of this sort thus has a potential storage time of seconds [13]. In conclusion all the abovementioned issues, such as linear and nonlinear scattering, recoil heating, intensity fluctuation, and pointing instabilities of the trapping beams, result in dipole-force fluctuations, and collisions with the background gas lead to heating up the qubit atom and therefore escaping from the trap [20]. This motivates the researchers to look for a way to create more stable quantum memories with longer storage time.

1.2 Rydberg atoms for quantum information

The optical degrees of freedom of single atoms such as polarization, wavelength, transverse mode, etc., can provide qubits or qudits for quantum information. Thus control and manipulation of single atoms are now of great interest given the potential to create quantum registers with single-atom techniques [21–24]. Furthermore, the observation of entanglement between a single atom and a single photon [25] provides the precondition for quantum communication and computation. Ground atomic states are ideal for preserving quantum coherence [26], but implementation of fast and deterministic quantum operations is challenging due to their weak interactions. Such considerations suggest to employ a quantum superposition of a ground and a Rydberg state to achieve both fast and deterministic quantum operations and long-lived memory [27]. A Rydberg atom is an atom in a highly excited state, typically with a principal quantum number $n \gg 1$. The excited valence electron can travel microns from the nucleus, while still remaining bound to it. Because the Rydberg electron is so far from the core of the atom, the Rydberg atom develops exaggerated properties, such as high polarizability. The strong dipolar interactions between Rydberg atoms can potentially be used for fast quantum gates between qubits stored in stable ground states of neutral atoms [28, 29]. The electric dipole strength of highly excited Rydberg atoms results in the Rydberg excitation blockade [28–33] which can be used in combination with electromagnetically induced transparency (EIT) [34, 35] to generate quantum states of light, entanglement of several atoms, and quantum logic gates [36–38]. The mapping of Rydberg interactions onto photons by means of EIT has emerged as a powerful approach to realizing few-photon optical nonlinearities [39–41] and provides the possibility to control the interaction between photons, which is a key ingredient to the goal of quantum information processing. Due to the nature of Rydberg blockade, the Rydberg nonlinearity is a sufficiently large and long range to build an optical quantum computer. The possibility to coherently control the quantum state of photon via dark-state Rydberg polaritons opens up interesting applications in reversible quantum memories for light waves [42] and shall find a possibility for a fast transformation [43].

1.3 Structured laser beam and atom

The development of structured laser beams such as Laguerre-Gaussian (LG) beams has enabled the coherent production of large number of identical spiral photons [44–47], which opens a new set of research in quantum optics in atomic level. LG beams are characterized by three quantum numbers: (i) the wave number k, (ii) the intertwined helical wave fronts l (an integer number) of azimuthal phase-dependent that features a screw dislocation, and (iii) the radial nodes p [48, 50]. Each spiral photon in the LG beams carries $l\hbar$ of intrinsic orbital angular momentum (OAM) along the direction of propagation [51], which is arising from their nonuniform spatial intensity distribution. The characteristic shape of the intensity distribution of LG beams as well as their intrinsic orbital angular momentum can be observed as a result of this interaction. The exchange of angular momentum induces a torque and an azimuthal shift in the resonant frequency besides the usual axial Doppler shift and recoil shift which tailored the control of the motion of the center of mass of the atoms to rotate about the beam axis [52, 53]. In higher order quadrupole transition processes, the internal motion of the atom participates in orbital angular momentum exchange between structured light and atom [54–57]. However, in dipole transition, the interaction of LG beam with atom and therefore transferring of OAM from LG beam to the external degree of freedom of the atom presents in the quantization of the center of mass motion of the atom [58].

A Rydberg atom with near classical size comparable to the wavelength of photon is big enough to feel phase differences of the helical wave front of LG beams. The LG beam has an advantage of control and narrows the resonances in electromagnetically induced transparency. The orbital angular momentum of LG beams has emerged as a popular choice for experiments on high-dimensional quantum information [59]. The dark center of blue-detuned Laguerre-Gaussian beam is emerged as an important feature to trap neutral atoms to decrease the atomic heating and decoherence rates and minimize the AC-Stark shift [60, 61]. In quantum information processing and data transmission, optical orbital angular momentum of spiral photons can provide an extra degree of entanglement [48] to carry quantum information in the different degrees of freedom by the higher dimensionality of the Hilbert space. In other words, the information carried by each photon can be increased significantly, from a qubit to a qudit, where d is the number of orthogonal basis vectors of the Hilbert space in which the photon lives.

1.4 Rydberg excitation

The excitation of a trapped neutral atom to a highly excited Rydberg state with long coherent time of ground-Rydberg transition is a promising platform for fast multi-qubit gates. The Rydberg state lifetime must be taken into account since the Rydberg state needs to be populated in order to implement two-qubit logical operations. The elementary operations necessary with Rydberg atoms in quantum information rely on the ability to coherently excite a Rydberg state and then returning them back to the ground state in controlled way so that the Rydberg atom is available for further processing. The lack of widely tunable frequency-stabilized ultraviolet wavelength diode lasers required for the direct excitation from the ground state to Rydberg states as well as weak direct excitation cross section has led to the use of multistep processes involving visible and near-infrared wavelengths [62–64]. To limit the spontaneous emission from the intermediate states in multistep excitation, which destroys the coherence, the population of the intermediate states can be manipulated by enlarging the detuning of the excitation laser frequency from the respective resonance frequency compared to the Rabi frequency of single-photon excitation. For large enough detunings, the intermediate states can be eliminated and a four-level system can be approximated to a two-level system with a total coupling strength. In this case, to obtain coherent coupling between the ground state and the Rydberg state, it is necessary to achieve an effective Rabi-frequency Ω well larger than the linewidth of the Rydberg state or of the driving laser fields. The coherent excitation of individual atoms trapped in tight optical dipole using two-photon excitation process has recently been described [65–68]. Also in order to cancel Doppler and recoil effects in Rydberg excitation, which limit the fidelity of the quantum gate [69], three-photon Rydberg excitation configuration is already proposed [70]. However, selectively localized and coherent excitation of atom to a highly excited Rydberg state and creation of a perfect blockade are still challenging [3, 71]. Firstly, the common red-detuned dipole traps can store atoms in the ground state with low decoherence, but they do not trap Rydberg states, and moreover, an atom in the Rydberg state moves in different optical potential than that experienced by the ground state. Secondly, the photoionization near the core due to the red-detuned trapping light, the sensitivity to the stray fields because of the large electron orbit and large polarizability that scales as $\propto n7$ [72], and the motional-induced dephasing presents crucial limits to the usable spectrum. Thirdly, although the intermediate state detuning reduces spontaneous emission, however, the small spontaneous emission besides of atomic motion and collisions [73] can limit storage times and therefore the ground-Rydberg atomic coherence time. Fourthly, the weak oscillator strength between ground and highly excited Rydberg state and very large electric dipole requires high laser excitation intensity to achieve the required signal-to-noise ratio. Fifthly, in most implementations using lattices or trap arrays generated with diffractive optics, it is relatively difficult to control the trap intensity and avoid perturbation of the atomic energy levels induced by inhomogeneous light distributions. Finally, in highly excited state, the energy separation is very small so that the selective excitation of the atom to a highly excited Rydberg state requires experimental technique with extremely high resolution.

To reduce errors in Rydberg excitation experiments, one solution is turning off the trapping system before the Rydberg excitation. However, it can cause problem for implementations with many qubits and creates unwanted Doppler shifts. This can be addressed by choosing a trap that works for both ground and Rydberg state of the atom. On the other hand, for the implementation of Rydberg-based quantum computing protocols with neutral atoms [3, 74–76], one can use a magic wavelength at which two atomic states (ground and Rydberg states) experience the same AC-Stark shift in a light field. In this chapter, a four-level Rydberg atom as a quantum memory in far-off resonance optical dipole-quadrupole trap is introduced. The controlling of the quantum state of localized Rydberg atom by tuning the excitation parameters is presented. It is shown that the ability to control the quantum state of Rydberg atom opens some interesting prospects for advances in quantum information processing.

2. High precision excitation and manipulation of a localized single atom to a highly excited Rydberg state

In this section, the theory of the three-step axial Doppler-free GGLG excitation of an alkali atom (e.g., Rb) from the ground state to the desired Rydberg state at the level of a single atomic excitation which is crucial for applications to quantum information processing is presented. The Rydberg excitation process is based on far-off resonance dipole-dipole-quadrupole transition. The quadrupole Rydberg transition in the last step is via LG-polarized laser beam. The Rydberg atom is localized at the dark center of structured beam, where the effect of atomic vibrations as well as AC-Stark shift is completely disappeared. The geometry of excitation as well as unique properties of LG-Rydberg excitation beam provides qudit of quantum memory in less disrupting effects such as motional heating effect, spontaneous emission due to high power, AC-Stark and Doppler, and recoil shift at the center of the trap which guarantee the high-fidelity gate.

2.1 Excitation configuration

The schematic diagram of a four-level atomic system is shown in **Figure 1**. We denote $|0>$ to be the atomic ground state and $|i>$ with $i = 1,2,3$ to be the respective first, second, and third excited states separated by energy E_i-E_{i-1}. The atom is cooled down to the recoil temperature and localized at the intersecting point of the excitation laser beams in the transversal plane z = 0 as sketched in **Figure 2(a)**. The wave vectors $\boldsymbol{k}_i$, $i = 1, 2, 3$, lie in x-z surface with the z axis directed along the LG-Rydberg excitation laser beams wave vector, $\boldsymbol{k}_3$. While the first and second Gaussian excitation laser beams wave vectors, $\boldsymbol{k}_1$ and $\boldsymbol{k}_2$, have the respective incidence angles θ_1 and

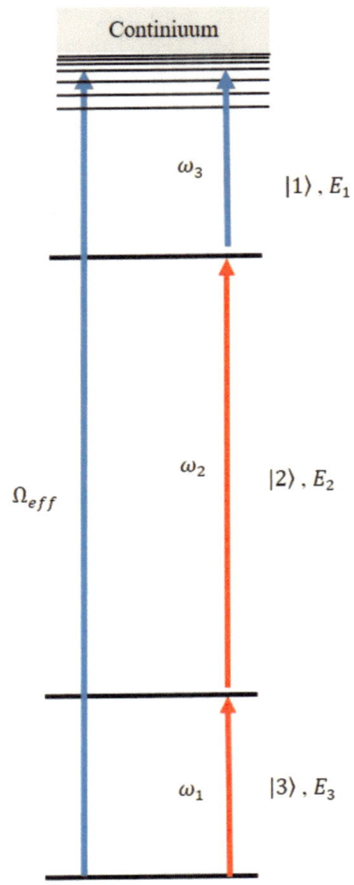

Figure 1.
Energy-level diagram three-photon Rydberg excitation [77].

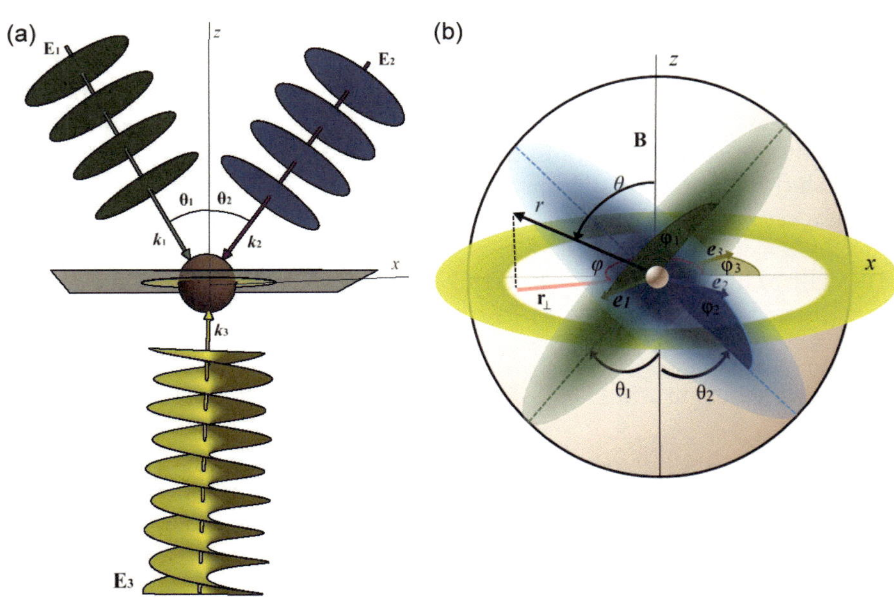

Figure 2.
(a) Geometry of Rydberg excitation via three-photon dipole-dipole-quadrupole transition. (b) Details of vectors and quantities involving in the excitation process [77].

θ_2 to the z axis. As it is explained in the following section, this special geometry provides the possibility for Doppler and recoil-free excitation.

The Rydberg atom-field interaction considered here corresponds to situations, where firstly the Rydberg atom's size is comparable to the LG-beam waist in the third Rydberg excitation and the Rydberg electron sees the variation of the light intensity. Different parts of the atom can feel different electric fields, and the quadrupole transition, which seems to be negligible in most situations due to a much stronger simultaneous effect of the dipole one, is considerably increased. Secondly, the Rydberg and two intermediate states decay by spontaneous emission, and it is assumed that the lifetime of the Rydberg state is much longer than the other states, and so its decay can be neglected during the time scales relevant to the excitation process. Finally, the dipole-dipole interaction between the Rydberg atoms, which may induce the blockade, is omitted to focus on the single-atom excitation mechanism.

2.2 Excitation laser beams

The four-level atom is driven by three laser fields, the Gaussian first and second excitation laser field with respective frequency ω_1 and ω_2 and the LG-Rydberg excitation laser field with frequency ω_3. Neglecting the focusing and the radial complexity and the mode curvature of the excitation laser beams, the electric field of each one-photon transition in the cylindrical-polar coordinates is given by

$$\mathbf{E}_{1,2}(R_\perp, \Phi, Z) = \mathbf{e}_{1,2} E_{01,2} f_{G1,2}(R_\perp) e^{i(\Theta_{1,2}(\mathbf{R}))}, \tag{1}$$

$$\mathbf{E}_{\ell p}(R_\perp, \Phi, Z) = \mathbf{e}_3 E_{03} \xi_{\ell p} \left(\frac{\sqrt{2}}{w_{03}}\right)^{|\ell|} f_{LG3}(R_\perp) e^{i(\Theta_3(\mathbf{R}))}, \tag{2}$$

where

$$f_{G1,2}(R_\perp) = e^{-\frac{R_\perp^2}{w_{01,2}^2}}, \tag{3}$$

$$f_{LG3}(R_\perp) = R_\perp^{|\ell|} e^{-\frac{R_\perp^2}{w_{03}^2}} L_p^{|\ell|}\left(\frac{2R_\perp^2}{w_{03}^2}\right), \tag{4}$$

$$\Theta_{1,2}(\mathbf{R}) = \mathbf{k} \cdot \mathbf{R}, \tag{5}$$

and

$$\Theta_3(\mathbf{R}) = \ell\varphi + \mathbf{k}_3 \cdot \mathbf{R}. \tag{6}$$

Here $\mathbf{e}_i$, w_{0i}, and $\mathbf{k}_i$ are the polarization, beam waist at z = 0, and wave vector of the ith excitation laser field, respectively. The electric field amplitudes E_{0i} are connected to the laser intensity I_i via $E_{0i} = \sqrt{(2I_i/c\varepsilon_0)}$, where c is the speed of light and ε_0 is the vacuum permittivity. In Eq. (2), ℓ corresponds to the optical orbital angular momentum, the mode index p represents the number of radial nodes of LG beam, $\xi_{\ell p} = \sqrt{\frac{2p!}{\pi(p+|\ell|)}}$ is the normalization constant, and $L_p^{|\ell|}$ is the Laguerre polynomial. The last excitation LG beam propagating along the z-direction in the plane of the focus of the beam appears as rings with radius R_l that is scaled linearly to the optical angular momentum and proportional to the beam waist [78]. The polarization $\mathbf{e}_i$ determines the particular transition conditions happened in ith laser beam. The appearance of the polarization vector $\mathbf{e}_i$ depends explicitly upon the choice of coordinate orientation. In the basis of $\mathbf{e}_\sigma$, where σ = {1, −1, 0} corresponds to the right circular, left circular, and linear polarization vector attached to the quantization axis z in xyz frame, the $\mathbf{e}_i$ are given by

$$\mathbf{e}_i(\theta_i, \varphi_i) = \sum_\sigma s_{i\sigma}(\theta_i, \varphi_i)\mathbf{e}_\sigma, \tag{7}$$

where $s_{i\sigma}(\theta_i, \varphi_i)$ is the relation coefficient and φ_i is the azimuthal angle of respective polarization according to **Figure 2(b)**.

The ith laser excitation beam carries linear momentum $\hbar \boldsymbol{k}_i$, and spin angular momentum $\pm\hbar$ per photon relates to the excitation laser polarization, if circularly polarized, while the last LG beam with an azimuthal phase dependence exp $(il\phi)$ in addition to linear and spin angular momentum carries orbital angular momentum that can be many times greater than the spin [49]. The combination of energy selectivity, associated with the laser light frequency, and sublevel selectivity, associated with polarization as well as angular momentum of light beam, provides good controls and manipulation over the qudits in quantum information processing.

2.3 The interaction of a four-level atom with three excitation laser fields

Let us begin with the four-level atom introduced in **Figure 1**. In general, the state of the atom for the coupled atom-laser system can be written in terms of the eigenstates of H_{atom}, the unperturbed Hamiltonian for the atomic system in the absence of the excitation laser, as

$$|\psi(\mathbf{r}, t)\rangle = c_0(t)\psi_0(\mathbf{r})e^{-i\frac{E_0}{\hbar}t}|0\rangle + \sum_i c_i(t)\psi_i(\mathbf{r})e^{-i\frac{E_i}{\hbar}t}|i\rangle, \tag{8}$$

where c_i (i = 0, 1, 2, 3) is the probability amplitude of ith state. The Hamiltonian for the system can be taken as

$$\widehat{H} = \widehat{H}_{atom} + \widehat{H}_{\text{int}}, \tag{9}$$

where $\widehat{H}_{\text{int}} = \widehat{H}_d + \widehat{H}_q$,

where $\mathbf{E}(R_\perp, \Phi, Z) = \mathbf{E}_1(R_\perp, \Phi, Z) + \mathbf{E}_1(R_\perp, \Phi, Z) + \mathbf{E}_{3lp}(R_\perp, \Phi, Z)$ is the total electric field of excitation lasers, and where $\widehat{H}_d$ and $\widehat{H}_q$ are the coupling to the electric dipole and quadrupole moment, respectively [79]. The first two excitations proceed via dipole transition employing the first two Gaussian laser beams, while the Rydberg excitation takes place through the quadrupole transition using the polarized LG beam.

To study the dynamics of the population of the different levels of the atom, we need to know c_i (t) by solving the Schrodinger equation:

$$i\hbar\frac{\partial}{\partial t}|\psi(\mathbf{r}, t)\rangle = \widehat{H}'|\psi(\mathbf{r}, t)\rangle, \tag{10}$$

where

$$\widehat{H}' = \widehat{U}\widehat{H}\widehat{U}^{-1} + -i\hbar\,\widehat{U}\frac{\partial}{\partial t}\widehat{U}^{-1}. \tag{11}$$

In matrix notation, the Schrodinger equation after rotating wave approximation can be given by

$$\frac{\partial}{\partial t}\begin{pmatrix} c_0(t) \\ c_1(t)e^{i\Delta_1 t} \\ c_2(t)e^{i(\Delta_2+\Delta_1)t} \\ c_3(t)e^{i\Delta t} \end{pmatrix} = -i\begin{pmatrix} 0 & \frac{\Omega_1}{2} & 0 & 0 \\ \frac{\Omega_1}{2} & \Delta_1 & \frac{\Omega_2}{2} & 0 \\ 0 & \frac{\Omega_2}{2} & \Delta_1+\Delta_2 & \frac{\Omega_3}{2} \\ 0 & 0 & \frac{\Omega_3}{2} & \Delta \end{pmatrix}\begin{pmatrix} c_0(t) \\ c_1(t)e^{i\Delta_1 t} \\ c_2(t)e^{i(\Delta_2+\Delta_1)t} \\ c_3(t)e^{i\Delta t} \end{pmatrix}. \tag{12}$$

Here

$$\Omega_{1,2}(R_\perp)=\left(\frac{2e^2 I_{1,2}}{\hbar^2 c\varepsilon_0}\right)^{\frac{1}{2}} \mathbf{r}_{0,1\to 1,2}\cdot\mathbf{e}_{1,2}\left(\theta_{1,2},\varphi_{1,2}\right) f_{G1,2}(R_\perp) \tag{13}$$

and

$$\Omega_3(R_\perp) = \left(\frac{2e^2 I_3}{\hbar^2 c\varepsilon_0}\right)^{\frac{1}{2}} \frac{2r_\perp \mathbf{r}_{2\to 3}}{\sqrt{\pi}w_{03}}\cdot \mathbf{e}_3(\theta_3,\varphi_3)e^{i\varphi}\frac{d}{dR_\perp}f_{LG3}(R_\perp) \tag{14}$$

are the first and second dipole and the last quadrupole Rydberg excitation Rabi frequencies, respectively, where $\mathbf{r}_{i-1\to i} = \langle i-1|\psi^*_{i-1}(\mathbf{r})\mathbf{r}\psi_i(\mathbf{r})|i\rangle$ stand for the atomic dipole momentums between states $i-1$ and i, with respective eigenfunctions $\psi_{i-1}(\mathrm{r})$ and $\psi_i(\mathrm{r})$; $\Delta_i = \omega_i - \frac{E_i - E_{i-1}}{\hbar} + \Delta_{iD}$ is the detuning of state i affected by residual Doppler shifts:

$$\Delta_{iD} = \Theta_i.\mathbf{v}_i \tag{15}$$

due to the random very small displacements and vibrations of the atom (see Eqs. (3) and (4)), even though the atom is cooled and trapped at the center of the LG beam, and $\Delta = \sum_{i=1}^{3}\Delta_i = E_3 - E_0 - \sum_i \omega_i$ stands for the total detuning to the Rydberg state.

According to Eq. (14), the quadrupole Rabi frequency of the Rydberg excitation step in addition to the polarization depends on the orbital angular momentum of the LG beam. The phase factor $e^{i\varphi}$ changes the parity symmetry and transfers one unit of optical orbital angular momentum to the internal motion of atom. The quadratic radial, $r_\perp \mathbf{r}_{2\to 3}$, in Eq. (14) is the consequence of the transversal variation of the LG-beam intensity.

2.3.1 Doppler and recoil shift compensation

The axial Doppler shift of the first, second, and Rydberg states by inserting Eqs. (3) and (4) in Eq. (14) can be, respectively, given as

$$\begin{aligned} \Delta_{1D} &= -k_1 v_1 \cos\theta_1 + k_1 v_1 \sin\theta_1, \\ \Delta_{2D} &= -k_2 v_2 \cos\theta_2 - k_2 v_2 \sin\theta_2, \\ \Delta_{3D} &= k_3 v_3 + \Delta_{LG}. \end{aligned} \tag{16}$$

It is noticeable that Δ_{3D} contains the azimuthal Doppler shift [80], $\Delta_{LG} = \frac{v_\Phi}{R_\perp}$, in addition to the axial Doppler shift, where v_Φ is the azimuthal velocity of the atom.

Doppler broadening due to atomic motion leads to imperfect Rydberg excitation which limits the fidelity of the entanglement that is created using Rydberg interactions. By adjusting θ_i, it is possible to get rid of the recoil and axial Doppler shift. According to **Figure 2(a)**, the excitation is recoil-free when

$$k_1 \sin\theta_1 = k_2 \sin\theta_2. \tag{17}$$

By substituting Eqs. (5) and (6) into Eq. (16), it is found that this flexible geometry of the excitation system results in the axial Doppler-free excitation condition in which the total detuning Δ is independent from the atomic vibrations but not rotations.

2.3.2 Adiabatic approximation and effective two-level transition

With the presence of decoherency, the evolution of the atomic states in memoryless environment will be expressed as density matrix formalism [81]. However, storing quantum information for long periods needs a decoherence-free quantum memory. In order to have coherent excitation, the spontaneous photon scattering should be limited by far detuning of laser excitation frequency from the respective excited state. In the limit of very large intermediate detuning such that, $|\Delta_i| \gg \Omega_i$ and $|\Delta_i| \gg |\Delta|$, the population of intermediate states is very low, and the system can behave as a two-level system with an effective coupling between ground and Rydberg states [77]. Supposing that the atom is initially in the ground state |0>, the time dependence of the Rydberg state population by the GGLG-beam excitation scheme can be obtained from Eq. (12) as

$$|c_3|^2 = \frac{\Omega_{eff}{}^2}{\Omega_{eff}{}^2 + \Delta_{eff}{}^2} \sin^2\left(\sqrt{\Omega_{eff}{}^2 + \Delta_{eff}{}^2}t\right), \tag{18}$$

where

$$\Delta_{eff}(R_\perp) = \Delta + \frac{\Omega_3^2(R_\perp)}{4\Delta_3} + \frac{\Omega_1^2(R_\perp)}{4\Delta_1} + \Delta_{LG}, \tag{19}$$

and

$$\Omega_{eff}(R_\perp) = \frac{\Omega_1(R_\perp)\Omega_2(R_\perp)\Omega_3(R_\perp)}{4\Delta_1\Delta_3} \tag{20}$$

are the effective detuning and resonant Rabi frequency characterizing effective quadrupole coupling between ground and Rydberg states due to the dipole-dipole-quadrupole transition under the adiabatic approximation.

The transverse profile of three-photon GGLG effective quadrupole Rabi frequency has a narrow central peak compared to the one-photon quadrupole LG Rabi frequency without any sidelope. Then, according to Eq. (18), the nonzero excitation probability is limited to the position of a single atom which is localized at the center of the trap leading to enhancement of the high-accuracy single-atom excitation.

2.3.3 Transition selection rules

To coherently excite a single atom to a Rydberg state for quantum information processing, it is crucial to determine the strength of the radial and angular momentum couplings between the ground and Rydberg states. Moreover, the quantum qudit encoding is based on the spin and orbital angular momentum exchange in the interaction of atom with excitation laser beams, which is expressed in angular momentum coupling term. The internal atomic wave function in spherical polar coordinate is written as

$$\psi_{nlm}(\mathbf{r}) = u_{nl}(r)Y_{lm}(\theta, \varphi), \tag{21}$$

where $u_{nl}(\mathrm{r})$ is the radial part of the electron wave function, which for a Rydberg state can be approximated using quantum defect theory [82], and n, l, and m are quantum numbers, which characterize the atom states. Substituting Eq. (21) into Eqs. (13) and (14) and considering the polarization orientation with respect to the quantization axis, the Rabi frequency of each transition can be given by

$$\Omega_1(0) = \left(\frac{8\pi I_1}{3\hbar^2\varepsilon_0 c}\right)^{\frac{1}{2}} eR_{n_0l_0\to n_1l_1}(r)\textstyle\sum_\sigma s_{1\sigma}(\theta_1,\varphi_1)\displaystyle\int_0^{2\pi}\int_0^{\pi} \sin\theta d\theta d\varphi \mathrm{Y}_{l_0}^{m_0} * \mathrm{Y}_1^{\sigma}\mathrm{Y}_{l_1}^{m_1}$$
$$= \left(\frac{2I_1}{\hbar^2\varepsilon_0 c}\right)^{\frac{1}{2}} eR_{n_0l_0\to n_1l_1}(r)\beta_{1dp} \tag{22}$$

$$\Omega_2(0) = \left(\frac{8\pi I_2}{3\hbar^2\varepsilon_0 c}\right)^{\frac{1}{2}} eR_{n_1l_1\to n_2l_2}(r)\textstyle\sum_\sigma s_{2\sigma}(\theta_2,\varphi_2)\displaystyle\int_0^{2\pi}\int_0^{\pi} \sin\theta d\theta d\varphi \mathrm{Y}_{l_1}^{m_1} * \mathrm{Y}_1^{\sigma}\mathrm{Y}_{l_2}^{m_2}$$
$$= \left(\frac{2I_2}{\hbar^2\varepsilon_0 c}\right)^{\frac{1}{2}} eR_{n_1l_1\to n_2l_2}(r)\beta_{2dp} \tag{23}$$

$$\Omega_3(0) = \left(\frac{8\pi I_3}{3\hbar^2\varepsilon_0 c}\right)^{\frac{1}{2}} \sqrt{\frac{32}{3}\frac{eR_{n_2l_2\to n_3l_3}(r)}{w_3}}\textstyle\sum_\sigma s_{3\sigma}(\theta_3,\varphi_3)\displaystyle\int_0^{2\pi}\int_0^{\pi} \sin\theta d\theta d\varphi \mathrm{Y}_{l_2}^{m_2} * \mathrm{Y}_1^{\sigma}\mathrm{Y}_1^{1}\mathrm{Y}_{l_3}^{m_3}$$
$$= \left(\frac{16I_3}{\pi\hbar^2\varepsilon_0 c}\right)^{\frac{1}{2}} \frac{eR_{n_2l_2\to n_3l_3}(r)}{w_{03}}\beta_{3qp} \tag{24}$$

where $R_{ni-1}, l_{i-1} \to n_i, l_i$ represent the overlap integral between the radial wave functions of the electron and the dipole-quadrupole moment and $\beta_{l_{i-1}m_{i-1}\to l_im_i}$ are the angular coupling expressed in terms of Clebsch-Gordan coefficients [77]. $\beta_{l_{i-1}m_{i-1}\to l_im_i}$ contributes in precise quadrupole Rydberg excitation with elemental parameters θ_i and Φ_i. Different combinations of θ_i and Φ_i provide all possible transitions accessible, which is applicable in precision measurements [83]. In case of $\theta_3 = 0$ as shown in **Figure 2(a)**, for left and right circularly polarized LG beam with p = 0 and ℓ = 1, the angular momentum transferred to the internal motion of the Rydberg atom via quadrupole transition is $|\Delta m| = 2$ and 0, respectively. Therefore, there is a possibility for the Rydberg atom to gain two units of angular momentum due to the quadrupole LG excitation: one from the polarization and the other from optical orbital angular momentum of LG beam. Additionally, the radial overlap integral of the quadrupole transition $R_{n_2l_2\to n_3l_3}(r)$ is considerable for high-lying Rydberg state with respect to w_{03}, which increases the quadrupole Rabi frequency. Moreover, the w_{03}-dependence of the Rydberg excitation Rabi frequency reflects the fact that the electric quadrupole transitions for an LG beam scales with w_{03} (compared to the plane wave, which scales with wave number k). Therefore, a relevant focusing with respect to the diffraction limit in addition to sufficient LG-beam power enhances the probability of the effective quadrupole excitation.

2.3.4 Localized Rydberg excitation in dipole-quadrupole potential landscape

Considering red- and blue-far-off resonance detuned first, second, and Rydberg excitation laser beams, respectively, due to the position-dependent AC-Stark shift of ground and Rydberg states as derived in Eq. (19), a position-dependent dipole-quadrupole potential landscape is found:

$$U_{FORDQT}(R_\perp) = U_{LG3} + U_{G1}, \tag{25}$$

where

$$U_{qLG3} = U_{03}\left(\frac{d}{dR_\perp} f_{LG3}(R_\perp)\right)^2, \tag{26}$$

and

$$U_{dG1} = -U_{01} f_{G1}^2(R_\perp)) \tag{27}$$

are the optical quadrupole and dipole potentials, respectively. In these expressions $U_{03} = \frac{\hbar}{4}\frac{|\Omega_3(0)|^2}{|\Delta_3(\mathbf{R},\mathbf{V})|}$ and $U_{01} = \frac{\hbar}{4}\frac{|\Omega_1(0)|^2}{|\Delta_1(\mathbf{R},\mathbf{V})|}$ are quadrupole and dipole potential depth. The first and the last excitation parameters are contributing in the self-trapping potential landscape called far-off resonance optical dipole-quadrupole trap (FORDQT), while the second deriving laser with high intensity increases the excitation probability. Turning the trap off also results in mechanical heating and decoherence due to entanglement between the qubit state and the center of mass motion. The far detuning from all atomic resonances substantially reduces the photon scattering rate, and the atom is localized in an almost conservative potential. It can be seen that the magnitude and direction of the force exerted on the atom depend on both the magnitude and sign of the intensity gradient and the detuning which pushes the atom back toward the dark center of the trap. The flexible geometry of the excitation configuration results in the axial Doppler- and recoil-free excitation at the center of the trap. The localized Rydberg excitation in FORDQT potential can pave the way to establish a new record for the length of the time that quantum information can be stored in and retrieved from a localized trapped Rydberg atom.

2.3.5 Optimization of the system by tuning the excitation parameters

The probability of coherent writing and reading a quantum state into and out of Rydberg atom as a long-lived memory depends on coherence time and strength of coupling between ground and Rydberg state. Substituting the Rabi frequencies defined in Eqs. (22)–(24) into Eqs. (19), (20) and (25), Δ_{eff}, Ω_{eff}, and U_{FORDQT} can be interpreted in terms of key parameters of excitation: the orientation of the excitation beams with respect to the quantum axis (θ_i), the intensity of laser excitation beams (I_i), the detuning from intermediate states (Δ_i), the laser polarization azimuthal angle (φ_i), and the orbital angular momentum and the beam waist of the Rydberg excitation LG beam. By adjusting θ_i the geometry of excitation can be constructed for a Doppler- and recoil-free excitation. The clever choice of φ_i results in the excitation to a desirable state. While Δ_i fulfills the far-off resonance condition, the intensity I_i can be adjusted to boost the effective quadrupole excitation with less effect of AC-Stark shift. Finally, the proper choice of excitation laser beam waists can result in a great reduction of trapping size. In the GGLG excitation

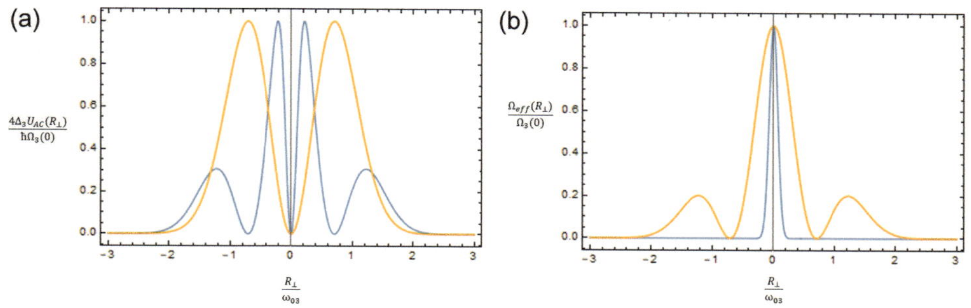

Figure 3.
(a) The far-off resonance optical dipole-quadrupole trapping potential normalized to the quadrupole potential versus transversal position of center of mass of the Rydberg atom, for the case $\Omega_1 = \Omega_3$, $|\Delta_1| = |\Delta_3|$, and $\omega_{01} = \omega_{02} = 0.2\ \omega_{03}$ (blue line) [77].

process, the Rydberg atom confines in the FORDQT potential, which is sensitive to the Rydberg atom position with respect to the transversal variations of the intensity of the excitation lasers which keeps the Rydberg atom at the minimum noise position of the trapping center and thus controls localized qudit state in longer coherent time reasonable for quantum information processing. Comparing to the Rydberg dipole excitation via LG beam, the Rydberg GGLG excitation system localizes the atom in a much smaller region. According to **Figure 3**, if the LG beam is focused into some micron waist, then in this self-trapping excitation system, the atom can be excited to a Rydberg state while tightly confined and controlled to sub-micron dimensions. Consequently, in a high-dimensional quantum information experiments via Rydberg excitation, care should be taken to relative control over all these parameters.

3. Summary

In this chapter, the Doppler- and recoil-free three-photon GGLG excitation promises to extend a single localized atom to a highly excited Rydberg state, which has application in the control and transformation of high-dimensional quantum states [16]. The adiabatic approximation results in an effective quadrupole Rabi frequency with a rich geometrical dependency. The quadrupole interaction in the last step of the LG excitation transfers a unit of orbital angular momentum to the Rydberg state in addition to the spin angular momentum. The GGLG excitation system allows to greatly reducing the rate of photon scattering and suppresses the loss rate due to collision, while the trapping potential of FORODQT localizes the Rydberg excitation and increases excitation coherency and allows for high detection efficiency and long detection time. A wide range of properties characterizing the excitation configuration can be controlled in real time through changes in geometry, polarization and orbital angular momentum, focal spot size, intensity, and frequency of the laser excitation beams to provide the ability to encode qudit in the internal degree of freedom of Rydberg atom independently to the center-of-mass motion. This aspect is vital to store and manipulate the quantum state of qudits in high-dimensional quantum information processing.

Author details

Leila Mashhadi[1]* and Gholamreza Shayeganrad[2]

1 Independent Researcher, Basel, Switzerland

2 Department of Biomedical Engineering, University of Basel, Allschwil, Switzerland

*Address all correspondence to: leila.mashhadi@yahoo.com

References

[1] Lvovsky AI, Sanders BC, Tittel W. Optical quantum memory. Nature Photonics. 2009;**3**:706-714

[2] Bloch I. Quantum coherence and entanglement with ultracold atoms in optical lattices. Nature. 2008;**453**:1016

[3] Saffman M, Walker TG. Analysis of a quantum logic device based on dipole-dipole interactions of optically trapped Rydberg atoms. Physical Review A. 2005;**72**:022347

[4] Specht HP, Nölleke C, Reiserer A, Uphoff M, Figueroa E, Ritter S, et al. A single-atom quantum memory. Nature. 2011;**473**:190-193

[5] Negretti A, Treutlein P, Calarco T. Quantum computing implementations with neutral particles. Quantum Information Processing. 2011;**10**:721

[6] Weitenberg C, Kuhr S, Mølmer K, Sherson JF. Quantum computation architecture using optical tweezers. Physical Review A. 2011;**84**:032322

[7] Ritter S et al. An elementary quantum network of single atoms in optical cavities. Nature. 2012;**484**:195

[8] Hofmann J et al. Heralded entanglement between widely separated atoms. Science. 2012;**337**:72

[9] Duan LM, Monroe C. Colloquium: Quantum networks with trapped ions. Reviews of Modern Physics. 2010;**82**: 1209

[10] Chen S et al. Deterministic and storable single-photon source based on a quantum memory. Physical Review Letters. 2006;**97**:173004

[11] Chen Y-A et al. Memory-built-in quantum teleportation with photonic and atomic qubits. Nature Physics. 2008;**4**:103

[12] Grimm R et al. Optical dipole traps for neutral atoms. Advances in Atomic, Molecular, and Optical Physics. 2000; **42**:95

[13] Davidson N et al. Long atomic coherence times in an optical dipole trap. Physical Review Letters. 1995;**74**: 1311

[14] Lengwenus A, Kruse J, Volk M, Ertmer W, Birkl G. Coherent manipulation of atomic qubits in optical micropotentials. Applied Physics B. 2007;**86**:377

[15] Miller JD, Cline RA, Heinzen DJ. Far-off-resonance optical trapping of atoms. Physical Review A. 1993;**47**: R4567

[16] Schlosser N, Reymond G, Protsenko I, Grangier P, Schlosser N, Reymond G, et al. Sub-poissonian loading of single atoms in a microscopic dipole trap. Nature. 2001;**411**:1024

[17] Volz J, Weber M, Schlenk D, Rosenfeld W, Kurtsiefer C, Weinfurter H. An atom and a photon. Laser Physics. 2007;**17**:1007

[18] Puppe T, Schuster I, Grothe A, Kubanek A, Murr K, Pinkse PWH, et al. Trapping and observing single atoms in a blue-detuned intracavity dipole trap. Physical Review Letters. 2007;**99**: 013002

[19] Nelson KD, Li X, Weiss DS. Imaging single atoms in a three-dimensional array. Nature Physics. 2007;**3**:556

[20] Phoonthong P, Douglas P, Wickenbrock A, Renzoni F. Characterization of a state-insensitive dipole trap for cesium atoms. Physical Review A. 2010;**82**:013406

[21] Weber M, Volz J, Saucke K, Kurtsiefer C, Weinfurter H. Analysis of

a single-atom dipole trap. Physical Review A. 2006;**73**:043406

[22] Schlosser N, Reymond G, Protsenko I, Grangier P. Sub-poissonian loading of single atoms in a microscopi dipole trap. Nature. 2001;**411**:1024-1027

[23] Zoller P. Quantum optics: Tricks with a single photon. Nature. 2000;**404**: 340-341

[24] Khudaverdyan M, Alt W, Dotsenko I, Kampschulte T, Lenhard K, Rauschenbeutel A, et al. Controlled insertion and retrieval of atoms coupled to a high-finesse optical resonator. New Journal of Physics. 2008;**10**:073023

[25] Blinov BB, Moehring DL, Duan LM, Monroe C. Observation of entanglement between a single trapped atom and a single photon. Nature. 2004;**428**:153-157

[26] Dudin YO, Li L, Kuzmich A. Light storage on the minute scale. Physical Review A. 2013;**87**:031801

[27] Li L, Kuzmich A. Quantum memory with strong and controllable Rydberg-level interactions. Nature Communications. 2016;**7**:13618

[28] Saffman M, Walker TG, Mølmer K. Quantum information with Rydberg atoms. Reviews of Modern Physics. 2010;**82**:2313

[29] Jaksch D et al. Fast quantum gates for neutral atoms. Physical Review. 2000;**85**:2208

[30] Choi JH, Knuffman B, Leibisch TC, Reinhard A, Raithel G. Cold Rydberg atoms. Advances in Atomic, Molecular, and Optical Physics. 2007;**54**:131

[31] Jaksch D, Cirac JI, Zoller P, Rolston SL, Côté R, Lukin MD. Fast quantum gates for neutral atoms. Physical Review Letters. 2000;**85**:2208

[32] Heidemann UR, Bendkowsky V, Butscher B, Löw R, Santos L, Pfau T. Evidence for coherent collective Rydberg excitation in the strong blockade regime rolf. Physical Review Letters. 2007;**99**:163601

[33] Gaëtan A, Evellin C, Wolters J, Grangier P, Wilk T, Browaeys A. Analysis of the entanglement between two individual atoms using global Raman rotations. New Journal of Physics. 2010;**12**:065040

[34] Firstenberg O, Adams CS, Hofferberth S. Nonlinear quantum optics mediated by Rydberg interactions. Journal of Physics B: Atomic, Molecular and Optical Physics. 2016;**49**:152003

[35] Mohapatra AK et al. A giant electro-optic effect using polarizable dark states. Nature Physics. 2008;**4**:890

[36] Isenhower L, Saffman M, Mølmer K. Multibit Ck NOT quantum gates via Rydberg blockade. Quantum Information Processing. 2011;**10**:755

[37] Muller MM, Murphy M, Montangero S, Calarco T, Grangier P, Browaeys A. Implementation of an experimentally feasible controlled-phase gate on two blockaded Rydberg atoms. Physical Review A. 2014;**89**: 032334

[38] Das S, Grankin A, Iakoupov I, Brion E, Borregaard J, Boddeda R, et al. Photonic controlled-PHASE gates through Rydberg blockade in optical cavities. Physical Review A. 2016;**93**: 040303(R)

[39] Peyronel T, Firstenberg O, Liang Q, Hofferberth S, Gorshkov A, Pohl T, et al. Quantum nonlinear optics with single photons enabled by strongly interacting atoms. Nature. 2012;**488**:57

[40] Pritchard JD, Maxwell D, Gauguet A, Weatherill KJ, Jones MPA, Adams

CS. Cooperative atom-light interaction in a blockaded Rydberg ensemble. Physical Review Letters. 2010;**105**: 193603

[41] Pritchard JD,Weatherill KJ, Adams CS. Nonlinear quantum optics mediated by Rydberg interactions. In: Madison KW et al., editors. Annual Review of Cold Atoms and Molecules. Vol. 1. 2013. pp. 301-350

[42] Kozhekin AE, Mølmer K, Polzik E. Quantum memory for light. Physical Review A. 2000;**62**:033809

[43] Müller MM, Haakh HR, Calarco T, Koch CP, Henkel C. Prospects for fast Rydberg gates on an atom chip. Quantum Information Processing. 2011; **10**:771

[44] Beijersbergen M, Coerwinkel R, Kristensen M, Woerdman J. Helical-wavefront laser beams produced with a spiral phaseplate. Optics Communications. 1994;**112**:321

[45] Curtis JE, Koss BA, Grier DG. Molecular distributed sensors using dark soliton array trapping tools. Optics Communications. 2002;**207**:169

[46] Mirhosseini M, na Loaiza OSM, Chen C, Rodenburg B, Malik M, Boyd RW. Rapid generation of light beams carrying orbital angular momentum. Optics Express. 2013;**21**:30196

[47] Beijersbergen M, Allen L, van der Veen H, Woerdman J. Astigmatic laser mode converters and transfer of orbital angular momentum. Optics Communications. 1993;**96**:123

[48] Wang J et al. Terabit free-space data transmission employing orbital angular momentum multiplexing. Nature Photonics. 2012;**6**(7):488-496

[49] Allen L et al. Orbital angular-momentum of light and the transformation of Laguerre-Gaussian laser modes. Physical Review A. 1992; **45**(11):8185-8189

[50] Mashhadi L, Mehrafarin M. Paraxial propagation in amorphous optical media with screw dislocation. Journal of Optics. 2010;**12**:035703

[51] Berry M. Paraxial beams of spinning light. Proceedings of SPIE. 1998;**3487**:6

[52] Allen L et al. Atom dynamics in multiple Laguerre-Gaussian beams. Physical Review A. 1996;**54**(5): 4259-4270

[53] Jáuregui R. Rotational effects of twisted light on atoms beyond the paraxial approximation. Physical Review A. 2004;**70**:033415-033411-033419

[54] Babiker M et al. Orbital angular momentum exchange in the interaction of twisted light with molecules. Physical Review Letters;**89**(14):143601

[55] Van Enk JK. Selection rules and center of mass motion of ultra cold atoms. Quantum Opt. 1994;**6**:445-457

[56] Alexandrescu A et al. Electronic and centre of mass transitions driven by Laguerre-Gaussian beams. Journal of Optics B: Quantum and Semiclassical Optics. 2005;**7**:87-92

[57] Kumar Mondal P, Deb B, Majumder S. Angular momentum transfer in interaction of Laguerre-Gaussian beams with atoms and molecules. Physical Review A. 2014;**89**:063418

[58] Muthukrishnan A, Stroud CR Jr. Entanglement of internal and external angular momenta of a single atom. Journal of Optics B: Quantum and Semiclassical Optics. 2002;**4**:S73-S77

[59] Krenn M, Malik M, Erhard M, Zeilinger A. Orbital angular momentum of photons and the entanglement of Laguerre-Gaussian modes. Philosophical

Transactions of the Royal Society A. 2017;**375**(2087):20150442

[60] Isenhower L, Williams W, Dally A, Saffman M. Atom trapping in an interferometrically generated bottle beam trap. Optics Letters. 2009; **34**:1159

[61] Ng J, Lin Z, Chan CT. Theory of optical trapping by an optical vortex beam. Physical Review Letters. 2010; **104**:103601

[62] Miroshnychenko Y et al. Coherent excitation of a single atom to a Rydberg state. Physical Review A. 2010;**82**: 013405

[63] Fahey DP, Michael W. Noel excitation of Rydberg states in rubidium with near infrared diode laser. Optics Express. 2011;**19**:17002

[64] Johnson LAM, Majeed HO, Varcoe BTH. Applied Physics B: Lasers and Optics. 2012;**106**:257

[65] Fahey DP, Noel MW. Excitation of Rydberg states in rubidium with near infrared diode lasers. Optics Express. 2011;**19**:17002

[66] Sheng J, Chao Y, Kumar S, Fan H, Sedlacek J, Shaffer JP. Intracavity Rydberg-atom electromagnetically induced transparency using a high-finesse optical cavity. Physical Review A. 2017;**96**:033813

[67] Johnson LAM, Majeed HO, Sanguinetti B, Becker T, Varcoe BTH. Absolute frequency measurements of 85RbnF7/2 Rydberg states using purely optical detection. New Journal of Physics. 2010;**12**:063028

[68] Johnson TA, Urban E, Henage T, Isenhower L, Yavuz DD, Walker TG, et al. Rabi oscillations between ground and Rydberg states with dipole-dipole atomic interactions. Physical Review Letters. 2008;**100**:113003

[69] Keating T, Cook RL, Hankin AM, Jau Y-Y, Biedermann GW, Ivan H. Deutsch Robust quantum logic in neutral atoms via adiabatic Rydberg dressing. Physical Review A. 2015;**91**: 012337

[70] Ryabtsev II, Beterov II, Tretyakov DB, Entin VM, Yakshina EA, Rzhanov AV. Doppler- and recoil-free laser excitation of Rydberg states via three-photon transitions. Physical Review A. 2011;**84**:053409

[71] Baur S, Tiarks D, Rempe G, Dürr S. Single-photon switch based on Rydberg blockade. Physical Review Letters. 2014; **112**:073901

[72] Heckötter J, Freitag M, Fröhlich D, Aßmann M, Bayer M, Semina MA, et al. Scaling laws of Rydberg excitons. Physical Review B. 2017;**96**:125142

[73] Niederprüm T, Thomas O, Manthey T, Weber TM, Ott H. Giant cross section for molecular ion formation in ultracold Rydberg gases. Physical Review Letters. 2015;**115**:013003

[74] Safronova MS, Williams CJ, Clark CW. Frequency-dependent polarizabilities of alkali atoms from ultraviolet through infrared spectral regions. Physical Review A. 2003;**67**: 040303

[75] Piotrowicz MJ, Lichtman M, Maller K, Li G, Zhang S, Isenhower L, et al. Two-dimensional lattice of blue-detuned atom traps using a projected Gaussian beam array. Physical Review A. 2013;**88**:013420

[76] Zhang S, Robicheaux F, Saffman M. Magic-wavelength optical traps for Rydberg atoms. Physical Review A. 2011;**84**:043408

[77] Mashhadi L. Three-photon Gaussian–Gaussian–Laguerre–Gaussian excitation of a localized atom to a highly excited Rydberg state. Journal of Physics B: Atomic, Molecular and Optical Physics. 2017;**50**:245201

[78] Curtis JE, Grier DG. Structure of optical vortices. Physical Review Letters. 2003;**90**:133901

[79] Lembessis VE, Babiker M. Enhanced quadrupole effects for atoms in optical vortices. Physical Review Letters. 2013; **110**:083002

[80] Allen L, Babiker M, Powerb WL. Azimuthal Doppler shift in light beams with orbital angular momentum. Optics Communication;**112**:141

[81] Blum K. Density Matrix Theory and Applications. New York: Plenum Press; 1981

[82] Seaton MJ. The quantum defect method. Monthly Notices of the Royal Astronomical Society. 1958;**118**:504

[83] Kreuter A, Becher C, GPT L, Mundt AB, Russo C, Häffner H, et al. Experimental and theoretical study of the 3d2D-level lifetime of 40ca+. Physical Review A. 2005;**71**:032504

Chapter 4

Single-Atom Field-Effect Transistor

Er'el Granot

Abstract

A simple single-atom transistor configuration is suggested. The transistor consists of only a nanowire, a single-point impurity (the atom), and an external capacitor. The transistor gate is controlled by applying a transverse voltage on the capacitor. The configuration does not rely on tunneling current and, therefore, is less sensitive to manufacturing processes since it requires less accuracy and fewer production processes. Moreover, unlike resonant-tunneling devices, the proposed transistor configuration does not suffer from a compromise between high speed and high extinction ratio. In fact, it is shown that this transistor can be extremely fast, without affecting the signal's extinction ratio, which can be as high as 100%.

Keywords: quantum dots, quantum point defect, point impurity, quantum transistor, single-atom transistor, field-effect transistor

1. Introduction

Despite the fact that the field-effect transistor (FET) was patented long before the formal invention of the transistor by Lilenfeld (in 1926) and Heil (in 1934), it was produced only two decades later when its patent expired. Nevertheless, its benefits were soon realized, and it became the building block of every integrated chip.

In 1975, Gordon Moore made a bold statement, which he updated a decade later, that the number of transistors on an integrated chip doubles every couple of years [1, 2]. This Moore's law is surprisingly still valid. In fact, it seems that this is the only parameter, which keeps growing exponentially for five consecutive decades. A simple extrapolation of this trend reveals that within a decade, the size of the average transistor should be no larger than the dimensions of a single atom.

The idea to manufacture few atom-based electronic devices was first suggested by Richard Feynman, but it has become a reality only after the scanning tunneling microscope (STM) was invented, and manipulations of single atoms became feasible [3].

Recently there have been several attempts to fabricate nano-devices, which are based on several atoms and even on a single atom [4–9]. These devices can operate as single-atom transistors [10–13]. The main problem with these devices is that while the device's core is based on a single atom, the connectors are considerably larger, and consequently, it is extremely complicated to model the device since the models are spread over several length scales.

In order to simplify the model, the atom and the leads should both be presented in the simplest form possible.

That was the main motivation to create a model, in which the entire transistor is within the leads [14]. This configuration is in high agreement with the experiment of a single-atom transistor [10] and, at the same time, can be simulated by a relatively simple model. The solutions of this model can be expressed, with great accuracy, by analytical expressions.

However, since this configuration is based on quantum tunneling, the single atom is not directly connected to the conducting leads (for resonant tunneling via a point defect without the insulators, see [15, 16]). Such a device is very difficult to manufacture, since the atom has to be encapsulated by the surrounding (other) atoms; it has to be located with great accuracy, and due to the resonant nature of the device, the atom must be located exactly at the center of the device; otherwise, the device's efficiency exponentially decreases.

However, resonant tunneling is not essential to achieve fine control. For example, it has been shown that a single-point defect in a nanowire can be a perfect reflector for certain energies. Moreover, the point defect can cause a universal conduction reduction. At certain Fermi energies, the conductance drops at exactly $2e^2/h$ [17].

Since the energy level of the point defect's bound state can be modified, then a simple nanowire with a single defect (a single atom) can be used as a nanotransistor. This is a much simpler device, which can be produced in fewer production stages than resonant-tunneling devices.

However, to control the resonance energy of the point defect, an external electric field should be applied. The field affects the entire device and does not selectively influence only the defect. Therefore, there is a need for a complete model, which integrates the nanowire, the point defect, and the electric field.

The object of this chapter is to present such a model of a nanotransistor, which is governed not by resonant-tunneling process but by Fano anti-resonance [18], which is generated by the interaction between the point defect and the nanowire. In this transistor configuration, the FET's gate is controlled by an external electric field.

2. The model

The system is presented in **Figure 1**. The system consists of an infinite nanowire (in the longitudinal x-direction), whose width (in the transverse y-direction) is w, a point defect, whose distance from the boundary of the nanowire is $w\varepsilon$, and an external capacitor, whose voltage and charge can be controlled by a power source.

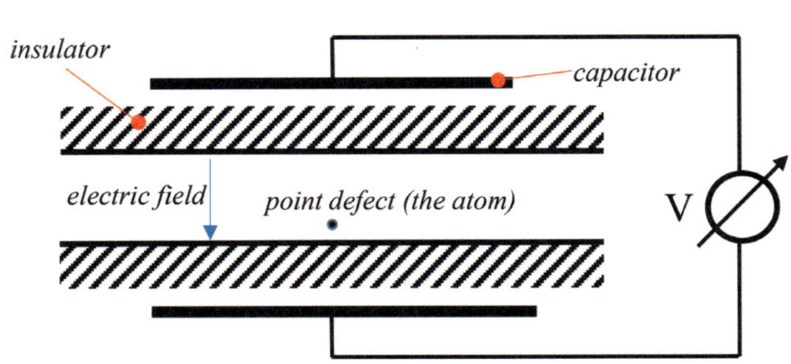

Figure 1.
Model schematic. The capacitor creates the transverse electric field, and the point defect simulates the single atom.

Mathematically, the model can be described by the 2D stationary Schrödinger equation

$$\frac{\partial^2\psi}{\partial x^2}+\frac{\partial^2\psi}{\partial y^2}+(E-U(y)+Fy)\psi=-D(\mathbf{r}-\mathbf{r}_0)\psi \tag{1}$$

in which normalized units (where Planck constant is $\hbar=1$ and the electron's mass is $m=1/2$) were used. In this equation

$$U(y)=\begin{cases}0 & 0<y<w\\ \infty & else\end{cases} \tag{2}$$

is the boundaries' potential, which confines the dynamics to the wire's geometry. F is the electric field.

The point defect, which models the single atom, is presented by the asymmetric impurity D function (IDF) [19–21]

$$D(\mathbf{r})\equiv\lim_{\rho\to0}\frac{2\sqrt{\pi}\delta(y)\exp\left(-x^2/\rho^2\right)}{\rho\ln\left(\rho_0/\rho\right)} \tag{3}$$

which is located at $\mathbf{r}_0=\hat{y}\varepsilon$, and ρ_0 is related to the impurity's bound eigenenergy by

$$E_0=-\frac{16\exp\left(-\gamma\right)}{\rho_0^2}\cong-\frac{8.98}{\rho_0^2} \tag{4}$$

where $\gamma\cong0.577$ is Euler constant [22]. On the other hand, the relation between ρ_0 and the physical properties of a real physical impurity (which has a finite radius a and a finite local potential V_0) is

$$\rho_0=2a\exp\left(\frac{2}{V_0a^2}+\frac{\gamma}{2}\right). \tag{5}$$

The homogeneous eigenstates solutions of the wire, i.e., solutions without the point defect, are

$$\psi_{n,E}(x,y)=\exp\left(ik_nx\right)\chi_n(y) \tag{6}$$

where $\chi_n(y)$ are the eigenstates of the one-dimensional differential equation

$$\frac{\partial^2\chi_n(y)}{\partial y^2}+(E_n-U(y)+Fy)\chi_n(y)=0 \tag{7}$$

with the corresponding eigenvalues

$$k_n^2=E-E_n. \tag{8}$$

These eigenstates can be written using the Airy functions Ai and Bi [22] and the normalized parameter

$$\xi\equiv yF^{1/3}+\left(E-k_n^2\right)/F^{2/3}=F^{1/3}\left\{y+\left(E-k_n^2\right)/F\right\}$$

as

$$\chi_n(\xi) = N\{Ai(\xi)Bi(\xi_0) - Ai(\xi_0)Bi(\xi)\} \tag{9}$$

where N is the normalization constant, $\xi_0 \equiv F^{1/3}\{E_n/F\}$, where the eigenvalues E_n are determined by the infinite solutions of the transcendental equation.

$\chi_n(\xi_w) = 0$, when $\xi_w \equiv F^{1/3}\{w + E_n/F\}$.

In the case of a weak electric field, the eigenstates can be written to a first order in the electric field as a superposition of the *free* (zero electric field) eigenstates

$$\phi_m(y) = \sqrt{\frac{2}{w}} \sin(m\pi y/w) \tag{10}$$

namely,

$$\chi_m(y) = \phi_m(y) + Fw^3 \sum_q \phi_q(y) \frac{(-1)^{m+q} - 1}{2} \frac{mq}{(m+q)^3(m-q)^3} \frac{8}{\pi^4} \tag{11}$$

with the corresponding eigenenergies (again to the first order in F)

$$E_n \cong (n\pi/w)^2 + \frac{1}{2}Fw. \tag{12}$$

Clearly, in the absence of the point defect (the atom), there is no coupling between the transverse direction and the longitudinal one, i.e., the capacitor cannot affect the longitudinal conductance.

There is an exception, of course, if the capacitor occupies a finite region in space, in which case the electric field does create a coupling between the modes. But in the regime of a weak electric field, this coupling a_{mq} is also very weak

$$a_{mq} = Fw^3 \frac{(-1)^{m+q} - 1}{2} \frac{mq}{(m+q)^3(m-q)^3} \frac{8}{\pi^4}. \tag{13}$$

Even to adjacent modes (where most of the energy is transferred), the coupling is very weak

$$a_{m,m\pm1} = \mp Fw^3 \frac{m(m\pm1)}{(2m\pm1)^3} \frac{8}{\pi^4} \tag{14}$$

For example, the coupling between the first and the second modes is as small as $a_{1,2} \cong 0.0061Fw^3$.

However, the presence of the point defect breaks the Cartesian symmetry and increases the coupling between the modes.

The general solution, which takes the point defect into account, is

$$\psi(\mathbf{r}) = \psi_{inc}(\mathbf{r}) - \frac{G^+(\mathbf{r},\mathbf{r}_0)\psi_{inc}(\mathbf{r}_0)}{1 + \int d\mathbf{r}' G^+(\mathbf{r}',\mathbf{r}_0)D(\mathbf{r}'-\mathbf{r}_0)} \int d\mathbf{r}' D(\mathbf{r}'-\mathbf{r}_0) \tag{15}$$

where $\psi_{inc}(\mathbf{r})$ is the incoming wavefunction, which can be chosen as one of the eigenmodes $\chi_n(y)$, which in the case of a weak electric field can be approximated by Eq. (11) (or Eq. (10), i.e., by $\phi_m(y)$); $G^+(\mathbf{r},\mathbf{r}_0)$ is the outgoing 2D Green function, i.e., $G^+(\mathbf{r},\mathbf{r}_0)$ is the solution of the partial differential equation

$$-\nabla^2 G^+(\mathbf{r},\mathbf{r}_0) + [U(y) - E - Fy]G^+(\mathbf{r},\mathbf{r}_0) = -\delta(\mathbf{r}-\mathbf{r}_0) \tag{16}$$

which can be written in terms of the 1D eigenstates $\chi_m(y)$ as

$$G(\mathbf{r},\mathbf{r}') = \sum_{n=1}^{\infty} \frac{\chi_n(y)\chi_n^*(y')}{2i\sqrt{E-E_n}} \exp\left(i\sqrt{E-E_n}|x-x'|\right). \tag{17}$$

The scattered solution is therefore

$$\psi(\mathbf{r}) = \exp(ik_n x)\chi_n(y) - \sum_{m=1}^{\infty} A_{n,m}\chi_m(y)\exp(ik_m|x|) \tag{18}$$

with the coefficients

$$A_{n,m} \equiv \lim_{\rho\to 0} \frac{\frac{\chi_n(\varepsilon)\chi_m^*(\varepsilon)}{2ik_m}}{\frac{\ln(\rho_0/\rho)}{2\pi} + \sum_{q=1}^{\infty}\frac{\chi_q(\varepsilon)\chi_q^*(\varepsilon)}{2ik_q}\exp\left(-|E-E_q|\rho^2/4\right)}. \tag{19}$$

The transmitted solution ($x > 0$) is thus

$$\psi(\mathbf{r}) = \sum_{m=1}^{\infty} \exp(ik_m x)\chi_m(y)\{\delta_{n,m} - A_{n,m}\}. \tag{20}$$

Eq. (18) is a generic solution; however, there are two types of energies, for which the solution reveals a universal pattern.

3. Universal transition patterns

When the particle's energy is equal exactly to one of transverse eigenenergies, i.e., when $E \to E_Q$, then the wavefunction reduces to a simple but *universal* expression

$$\psi(\mathbf{r}) = \sum_{m=1}^{\infty} \exp(ik_m x)\chi_m(y)\left\{\delta_{n,m} - \delta_{Q,m}\frac{\chi_n(\varepsilon)}{\chi_Q(\varepsilon)}\right\} = \exp\left(i\sqrt{E_Q-E_n}x\right)\chi_n(y) - \chi_Q(y)\frac{\chi_n(\varepsilon)}{\chi_Q(\varepsilon)}. \tag{21}$$

A similar universality was shown for zero-field wire [23] (for other patterns, see [24])

$$\psi(\mathbf{r}) = \exp\left(i\pi\sqrt{Q^2-n^2}x\right)\sin\left(\frac{n\pi y}{w}\right) - \sin\left(\frac{Q\pi y}{w}\right)\frac{\sin(n\pi\varepsilon/w)}{\sin(Q\pi\varepsilon/w)}, \tag{22}$$

but Eq. (21) solution is valid in the presence of an electric field as well.

This solution is universal in the sense that it is totally independent of the point defect's strength (potential), which is manifested by the parameter ρ_0—a parameter that does not appear in Eq. (21). This pattern is presented in the upper panel of **Figure 2**.

Clearly, when the defect is close to the surface, i.e., $\varepsilon/w << 1$, then the solution is even independent of ε

$$\psi(\mathbf{r}) \cong \exp\left(i\pi\sqrt{Q^2-n^2}x\right)\sin\left(\frac{n\pi y}{w}\right) - \sin\left(\frac{Q\pi y}{w}\right)\frac{n}{Q}. \tag{23}$$

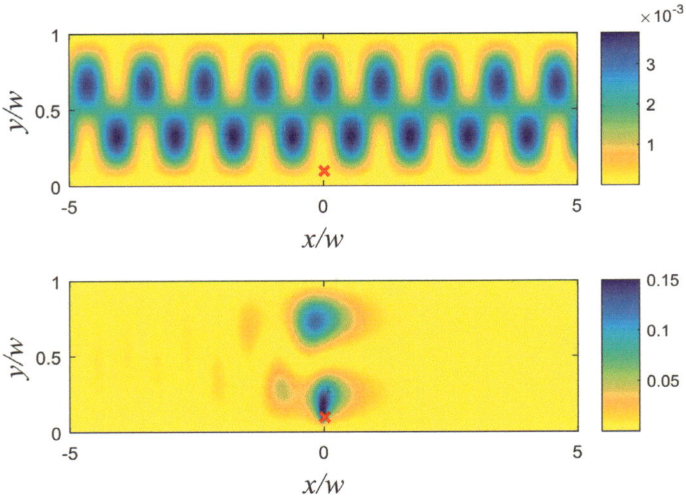

Figure 2.
A false color presentation of the conductance pattern. The spatial distribution in the wire of the probability density $|\psi(\mathbf{r})|^2$ is plotted at the transition energy $E = E_2 \cong (2\pi/w)^2 + Fw/2$ (upper panel) and at the zero-current energy $E = E_2^{\min}$ (lower panel). The crosses represent the point defect's location.

At this operation point, the transistor experiences maximum transmission with maximum current, which is universal and is independent of the point defect parameters. The defect deforms the conducting pattern, but it does not transfer any current to the mth mode. Consequently, the current remains in the initial nth mode, as if the defect is absent.

4. Universal conductance reduction

Another important case, which is going to be relevant to the transistor operation, occurs below the next energy transition where there is a dip in the transmission coefficient, and the conductance decreases by exactly $2e^2/h$.

Let the incoming particle energy be within the energy range $E_{Q-1} < E < E_Q$; for which case it is convenient to split the denominator of the coefficient, i.e., to rewrite Eq. (19) as

$$A_{n,m} \equiv \lim_{\rho\to 0} \frac{\frac{\chi_n(\varepsilon)\chi_m^*(\varepsilon)}{ik_m}}{\frac{\ln(\rho_0/\rho)}{\pi} + \sum_{q=1}^{Q-1} \frac{\chi_q(\varepsilon)\chi_q^*(\varepsilon)}{i\sqrt{E-E_q}} - \frac{\chi_Q(\varepsilon)\chi_Q^*(\varepsilon)}{\sqrt{E_Q-E}} - \sum_{q=Q+1}^{\infty} \frac{\chi_q(\varepsilon)\chi_q^*(\varepsilon)}{\sqrt{E_q-E}} \exp\left(-|E-E_q|\rho^2/4\right)} \tag{24}$$

or

$$A_{n,m} \equiv \frac{\frac{\chi_m^*(\varepsilon)\chi_n(\varepsilon)}{i\sqrt{E-E_m}}}{\frac{\ln(\rho_0/R(\varepsilon))}{\pi} + \sum_{q=1}^{Q-1} \frac{\chi_q(\varepsilon)\chi_q^*(\varepsilon)}{i\sqrt{E-E_q}} - \frac{\chi_Q(\varepsilon)\chi_Q^*(\varepsilon)}{\sqrt{E_Q-E}}}, \tag{25}$$

where

$$\ln R(\varepsilon) \equiv \lim_{\rho\to 0} \left[\ln\rho + \pi \sum_{q=Q+1}^{\infty} \frac{\chi_q(\varepsilon)\chi_q^*(\varepsilon)}{\sqrt{E_q-E}} \exp\left(-|E-E_q|\rho^2/4\right) \right]. \tag{26}$$

The device's conductance can be evaluated as ([25, 26])

$$G = \frac{1}{\pi} \sum_{m,\, l < n} T_{ml}, \tag{27}$$

where

$$T_{n,m} = |\delta_{n,m} - A_{n,m}|^2 \frac{k_m}{k_n} = \begin{cases} |1 - A_{n,n}|^2 & n = m \\ |A_{n,m}|^2 \dfrac{k_m}{k_n} & n \neq m \end{cases} \tag{28}$$

are the transmission coefficients (from the nth to the mth modes) of the wire with the impurity.

At the transition points, i.e., when $E = E_Q$ ($k_Q = 0$), the conductance is exactly

$$G = \frac{1}{\pi}(Q - 1), \tag{29}$$

which is $G = 2\frac{e^2}{h}(Q - 1)$ in ordinary physical units.

On the other hand, at the minimum transmission point ($E = E_Q^{\min} < E_Q$), when the real part of the denominator of Eq. (25) vanishes, i.e., when

$$\frac{\ln(\rho_0/R(\varepsilon))}{\pi} = \frac{\chi_Q(\varepsilon)\chi_Q^*(\varepsilon)}{\sqrt{E_Q - E_Q^{\min}}}, \tag{30}$$

then

$$A_{n,m} \equiv \frac{\frac{\chi_m^*(\varepsilon)\chi_n(\varepsilon)}{i\sqrt{E - E_m}}}{\sum_{q=1}^{Q-1} \frac{\chi_q(\varepsilon)\chi_q^*(\varepsilon)}{i\sqrt{E - E_q}}} = \frac{1}{\frac{\sqrt{E - E_m}}{\chi_m^*(\varepsilon)\chi_n(\varepsilon)} \sum_{q=1}^{Q-1} \frac{\chi_q(\varepsilon)\chi_q^*(\varepsilon)}{\sqrt{E - E_q}}} = \frac{1}{\frac{k_m}{\chi_m^*(\varepsilon)\chi_n(\varepsilon)} \sum_{q=1}^{Q-1} \frac{\chi_q(\varepsilon)\chi_q^*(\varepsilon)}{k_q}}. \tag{31}$$

Using the definition

$$\sigma \equiv \sum_{q=1}^{Q-1} \frac{\chi_q(\varepsilon)\chi_q^*(\varepsilon)}{\sqrt{E - E_q}} = \sum_{q=1}^{Q-1} \frac{\chi_q(\varepsilon)\chi_q^*(\varepsilon)}{k_q} \tag{32}$$

the conductance $G = \frac{1}{\pi}\sum_{n,\, m < Q} |\delta_{n,m} - A_{n,m}|^2 \frac{k_m}{k_n}$ is equal exactly to

$$G = \frac{1}{\pi} \sum_{n,\, m < Q} \left| \delta_{n,m} - \frac{\chi_m^*(\varepsilon)\chi_n(\varepsilon)}{k_m \sigma} \right|^2 \frac{k_m}{k_n} = \frac{1}{\pi} \sum_{n,\, m < Q} \left[\delta_{n,m} - 2\frac{\chi_m^*(\varepsilon)\chi_n(\varepsilon)}{k_m \sigma} + \left| \frac{\chi_m^*(\varepsilon)\chi_n(\varepsilon)}{k_m \sigma} \right|^2 \right] \frac{k_m}{k_n} = \frac{Q - 2}{\pi} \tag{33}$$

which is $G = 2\frac{e^2}{h}(Q - 2)$ in ordinary physical units.

Therefore, there is exactly a one unit of conductance reduction between the transition energy E_Q and the minimum point just below it $E_Q^{\min}$

$$\Delta G = G(E_Q) - G\left(E_Q^{\min}\right) = \pi^{-1}, \tag{34}$$

which is $2e^2/h$ in ordinary physical units.

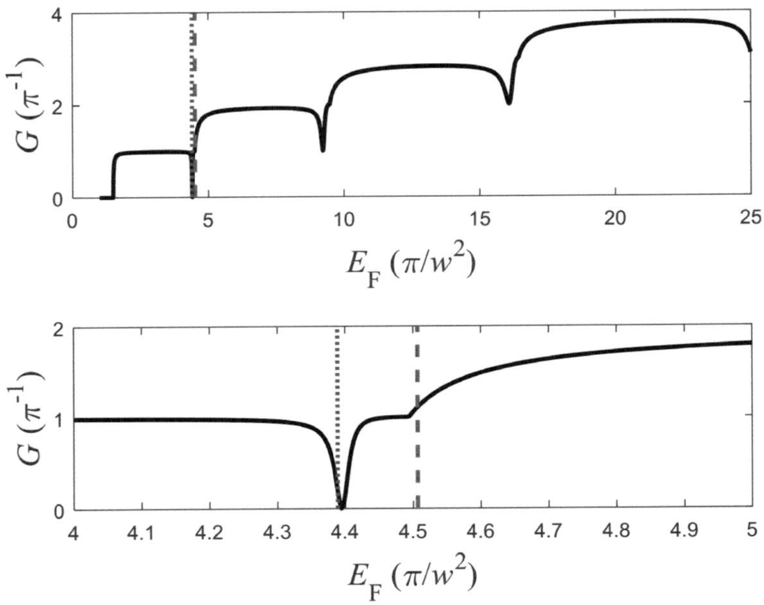

Figure 3.
The nanowire's conductance as a function of the particles' Fermi energy. The lower panel is a magnification of the transition zone. The dashed perpendicular line stands for the transition energy formula (Eq. (12)) $E = E_2$*, and the dotted line corresponds to the zero-current energy formula (Eq. (30))* $E = E_2^{\min}$.

This result is a generalization of [17]. The probability density at the point of minimum conductance is presented in the lower panel of **Figure 2**, and the dependence of the conductance on the particles' Fermi energy is presented in **Figure 3**. The minima are clearly seen.

Moreover, the approximate analytical expressions of the transition energy Eq. (12) and the minimum energy Eq.(30) are presented by horizontal lines.

5. Zero transmission point

The current can vanish only when the Fermi energy is within the energy range $\pi^2/w^2 < E_F < 4\pi^2/w^2$, in which case only the first mode is propagating. The transmission of the first mode is

$$t_{1,1} = 1 - \left\{ 1 + \frac{i\sqrt{E_2 - E_1}}{|\chi_1(\varepsilon)|^2} \left[\frac{\ln(\rho_0/R(\varepsilon))}{\pi} - \frac{|\chi_2(\varepsilon)|^2}{\sqrt{E_2 - E}} \right] \right\}^{-1}, \tag{35}$$

in which case the zero-current energy is approximately

$$E_R \cong E_2 - \frac{\pi^2 |\chi_2(\varepsilon)|^4}{\ln^2(\rho_0/R(\varepsilon))} \tag{36}$$

and in the case of weak fields, it can be written

$$t_{1,1} = 1 - \left\{ 1 + \frac{i\pi\sqrt{3}}{2\sin^2(\pi\varepsilon/w)} \left[\frac{\ln(\rho_0/R(\varepsilon))}{\pi} - \frac{2\sin^2(2\pi\varepsilon/w)}{w\sqrt{(2\pi/w)^2 + \frac{1}{2}Fw - E}} \right] \right\}^{-1} \tag{37}$$

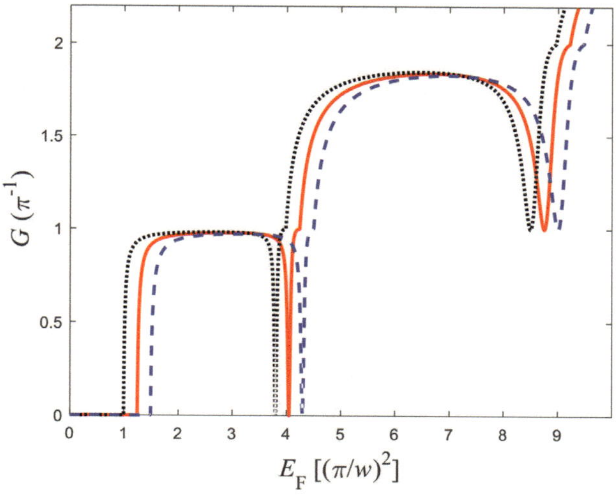

Figure 4.
The dotted curve stands for $Fw^3 = 0$*; the solid curve stands for* $Fw^3 = 5$*; and the dashed curve stands for* $Fw^3 = 10$.

with the zero-current (zero transmission) energy of

$$E_R \cong \left(\frac{2\pi}{w}\right)^2 + \frac{1}{2}Fw - \frac{4\pi^2 \sin^4(2\pi\varepsilon/w)}{w^2 \ln^2(\rho_0/R(\varepsilon))}. \tag{38}$$

In the case of a surface defect, i.e., when the atom is close to the wire's boundary (see **Appendix A**),

$$R(\varepsilon) \cong 4\varepsilon \exp(\gamma/2), \tag{39}$$

and then the zero-current energy is approximately

$$E_R \cong \left(\frac{2\pi}{w}\right)^2 + \frac{1}{2}Fw - \frac{\varepsilon^4 (2\pi/w)^6}{\ln^2(\rho_0 \exp(-\gamma/2)/4\varepsilon)}. \tag{40}$$

Therefore, the zero-current energy has a linear dependence on the electric field (and thus on the applied external voltage). In **Figure 4** this property is presented by plotting the conductance for three different transverse electric fields.

6. The transistor working point

In **Figure 5,** the conductance as a function of the normalized applied electric field is plotted. The transistor can work as a digital device, where the field varies between the binary cases:

$$T_{1,1} = \begin{cases} 0 & for \quad F = F_R \\ 1 & for \quad F = 0 \end{cases} \tag{41}$$

where

$$\frac{1}{2}F_R w \equiv E_F - \left(\frac{2\pi}{w}\right)^2 + \frac{4\pi^2 \sin^4(2\pi\varepsilon/w)}{w^2 \ln^2(\rho_0/R(\varepsilon))} \cong E_F - \left(\frac{2\pi}{w}\right)^2 + \frac{\varepsilon^4 (2\pi/w)^6}{\ln^2(\rho_0 \exp(-\gamma/2)/4\varepsilon)}. \tag{42}$$

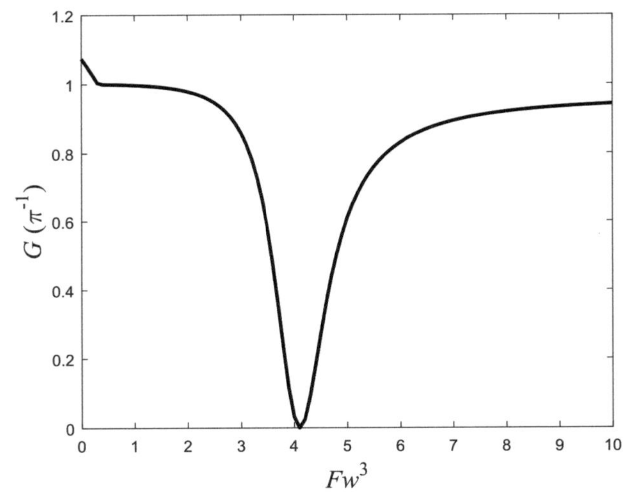

Figure 5.
Plot of the conductance as a function of the applied electric field for the particles' energy $E = 4(\pi/w)^2$.

But the transistor can also work in an analog mode as an amplifier, in which case the applied voltage should be modulated with respect to the bias voltage $F_* w$, which is the center of the linear regime, i.e.,

$$F_* w/2 = E_F - (2\pi/w)^2 + \delta^2 \left[1 \pm 2\frac{|\chi_1(\varepsilon)|^2}{|\chi_2(\varepsilon)|^2}\frac{\delta}{\sqrt{E_2 - E_1}}\right]. \tag{43}$$

Using this bias voltage, the transmission coefficient can be written

$$t_{1,1} = 1 - \left\{1 + ic\left[1 - \frac{\delta}{\sqrt{\delta^2/(1-c^{-1})^2 + (F - F_*)w/2}}\right]\right\}^{-1} \tag{44}$$

where
$c \equiv \frac{\sqrt{E_2 - E_1}}{\delta}\frac{|\chi_2(\varepsilon)|^2}{|\chi_1(\varepsilon)|^2} \cong 4\pi\sqrt{3}/\delta$, and $\delta \equiv \pi|\chi_2(\varepsilon)|^2/\ln(\rho_0/R(\varepsilon)) \cong 4\pi(2\pi\varepsilon)^2/$ $[w^3 \ln(\rho_0 \exp(-\gamma/2)/4\varepsilon)]$.

Therefore, the transistor gain at the working point is the ratio between the change in conductance and the applied transverse voltage $\Delta v = (F - F_*)w$, which is

$$gain = \frac{\Delta G}{\Delta v} = \frac{c(1-c^{-1})^3}{4\pi\delta^2} \tag{45}$$

when the point defect is a surface one, i.e., $\varepsilon < < 1$, then $\delta < < 1$ and $c >> 1$

$$gain \cong \frac{\sqrt{3}}{\delta^3} = \frac{\sqrt{3}w^9 \ln^3(\rho_0 \exp(-\gamma/2)/4\varepsilon)}{(4\pi)^3(2\pi\varepsilon)^6}, \tag{46}$$

which can be extremely large.

7. Fast switching

When the dip of the resonances is very narrow, the gain is very high; however, in this case, the transistor response is very slow, because it takes a substantial amount of time to establish the resonance. In fact, the gain is proportional to the transistor's time response τ, i.e.,

$$gain \propto \left(4\pi^2/w^2 - E\right)^{-1} \propto \tau. \tag{47}$$

However, the value of both can be controlled by changing the defect's parameters. Since

$$\rho_0 \cong 4\varepsilon \exp\left[\frac{\gamma}{2} + 2\pi \frac{\sin^2(2\pi\varepsilon/w)}{w\sqrt{4\pi^2/w^2 - E}}\right], \tag{48}$$

the parameter ρ_0 can be chosen to place the resonance dip at any point in the regime $\pi^2/w^2 < E < 4\pi^2/w^2$ and thus to determine the transistor time response

$$\rho_0 \cong 4\varepsilon \exp\left[\frac{\gamma}{2} + \frac{2\pi}{w}\sin^2(2\pi\varepsilon/w)\sqrt{\tau}\right]. \tag{49}$$

Therefore, the transistor with the quickest response is the one with a surface defect with

$$\rho_0 \cong 4\varepsilon \exp(\gamma/2). \tag{50}$$

In this case, the transistor time response is determined by the wire's width, i.e.,

$$\tau \sim (w/\pi)^2, \tag{51}$$

which in ordinary physical units is $\tau \sim m(w/h)^2$, i.e., the narrower the wire, the shorter the transistor's time response is.

Eq. (50) teaches that such a single-atom nanotransistor can be faster than any of the cutting-edge available transistors.

It should be emphasized that the point defect does not necessarily have to be an atom. It could be a molecule or any quantum dot that can be designed of having the necessary de-Broglie wavelength ρ_0.

8. Summary and conclusions

An innovative single-atom transistor configuration is suggested, which can be simplified and simulated by a simple model. The model consists of a narrow conducting wire, a single-point defect, and an electric field. This device's configuration does not require fine atomic-size gate contact and atomic-size accuracy for positioning the single atom. The device's mechanism is not based on resonant tunneling, and therefore, high accuracy is less essential. The gate is a capacitor that can be considerably larger than the point defect. Moreover, it was shown that this device can be extremely fast with a time response much shorter than any cutting-edge transistor.

The temporal analysis reveals a clear advantage of this configuration over resonant-tunneling ones (like [10, 14]). In resonant-tunneling devices, the signal's extinction ratio depends on the resonance state's lifetime. That is, there is no "zero-current" in resonant-tunneling devices. The minimum current ("zero") is actually a tunneling current, which is inversely proportional to the resonance state's lifetime. Therefore, in resonant-tunneling devices, "fast device" and "low minimum current" are competing demands. When seeking the former, one has to compromise on the latter, and vice versa.

No such compromise is required in the proposed transistor configuration since it has been shown that this configuration always keeps (at least theoretically) extinction ratio of 100%.

A. Appendix A: derivation of Eq. (39)

The expression (26), i.e.,

$$R(\varepsilon) = \exp \lim_{\rho\to 0}\left[\ln\rho + \pi\sum_{q=3}^{\infty}\frac{\left|\chi_q(\varepsilon)\right|^2}{\sqrt{E_q - E}}\exp\left(-\left|E - E_q\right|\rho^2/4\right)\right]$$

as a function of the defect's location ε is plotted for several energy values in the energy range $\pi^2/w^2 < E < 4\pi^2/w^2$ in **Figure A1**.

As can be seen from this figure, while there are considerably large variations around $\varepsilon \cong 0.2$, the differences in the value of $R(\varepsilon)$ for $\varepsilon \cong 0.5$, i.e., when the defect is located at the center of the wire, are relatively mild, and in which case $R(\varepsilon) \cong 0.3$. Moreover, in the case of a surface defect, i.e., $\varepsilon < < 1$, $R(\varepsilon)$ is independent of the particles' energy.

Using the definition $\xi \equiv \pi q\rho/w$ and the weak field approximation

$$\ln R(\varepsilon) \equiv \lim_{\rho\to 0}\left[\ln\rho + \frac{2}{w}\pi\exp\left(\pi^2\rho^2/w^2\right)\sum_{q=3}^{\infty}\rho\frac{\sin^2(\xi\varepsilon/\rho)}{\sqrt{\xi^2 - (4\pi\rho)^2/w^2}}\exp\left(-\xi^2/4\right)\right] \tag{A1}$$

which can be written as an integral

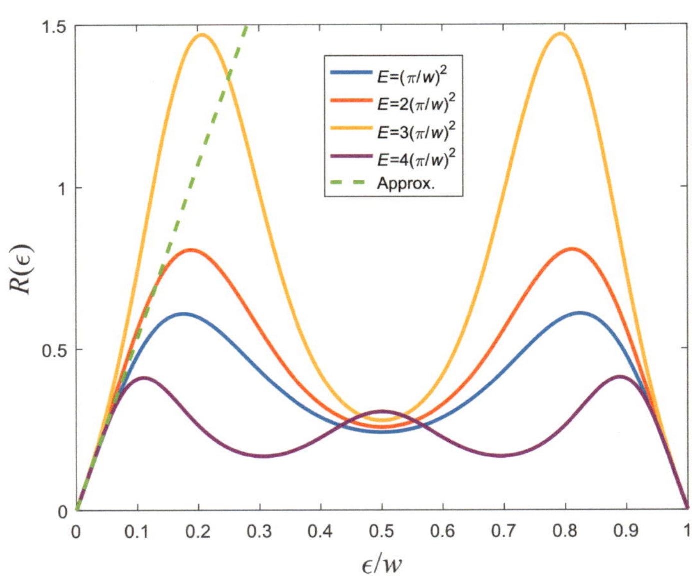

Figure A1.
Plots of $R(\varepsilon)$ as a function of the point defect's position in the wire ε for various energies. The dashed line is the small $\varepsilon/w < < 1$ approximation (Eq. (39)).

$$\ln R(\varepsilon) \equiv \lim_{\rho\to 0}\left[\ln\rho + 2\int_{\pi 3\rho}^{\infty}\frac{\sin^2(\xi\varepsilon/\rho)}{\sqrt{\xi^2-(4\pi\rho)^2/w^2}}\exp\left(-\xi^2/4\right)d\xi\right]. \tag{A2}$$

And due to the limit,

$$R(\varepsilon) \cong \exp\lim_{\rho\to 0}\left[\ln\rho + 2\int_{0}^{\infty}\frac{\sin^2(\xi\varepsilon/\rho)}{\xi}\exp\left(-\xi^2/4\right)d\xi\right]. \tag{A3}$$

Now, since for $\rho/w < < 1$

$$\ln(\rho/w) \cong \ln 2 - \gamma/2 - \int_{\rho/w}^{\infty}\frac{\exp\left(-\xi^2/4\right)}{\xi}d\xi, \tag{A4}$$

then

$$\begin{aligned} R(\varepsilon) &\cong \exp\lim_{\rho\to 0}\left[\ln 2w - \gamma/2 - \int_{\rho/w}^{\infty}\frac{\exp\left(-\xi^2/4\right)}{\xi}d\xi + 2\int_{0}^{\infty}\frac{\sin^2(\xi\varepsilon/\rho)}{\xi}\exp\left(-\xi^2/4\right)d\xi\right] = \\ &\exp\lim_{\rho\to 0}\left[\ln 2w - \gamma/2 - \int_{\rho/w}^{\infty}\frac{\cos(2\xi\varepsilon/\rho)}{\xi}\exp\left(-\xi^2/4\right)d\xi + 2\int_{0}^{\rho/w}\frac{\sin^2(\xi\varepsilon/\rho)}{\xi}\exp\left(-\xi^2/4\right)d\xi\right]. \end{aligned} \tag{A5}$$

Moreover, since, $\rho/\varepsilon \to 0$ but also $\varepsilon/w \to 0$, then in both integrals, the exponents can be ignored, i.e.,

$$R(\varepsilon) \cong \exp\lim_{\rho\to 0}\left[\ln 2w - \gamma/2 - \int_{\rho/w}^{\infty}\frac{\cos(2\xi\varepsilon/\rho)}{\xi}d\xi + 2\int_{0}^{\rho/w}\frac{\sin^2(\xi\varepsilon/\rho)}{\xi}d\xi\right] \tag{A6}$$

which finally yields the analytical expression

$$R(\varepsilon) \cong \exp\left[\ln 2w - \gamma/2 + \mathrm{Ci}(2\varepsilon/w) + \varepsilon^2/w^2\right] = 4\varepsilon\exp\left(\frac{\gamma}{2}\right), \tag{A7}$$

where Ci(x) is the cosine integral function [22].

Author details

Er'el Granot
Department of Electrical and Electronics Engineering, Ariel University, Ariel, Israel

*Address all correspondence to: erel@ariel.ac.il

References

[1] Moore G. Chapter 7: Moore's law at 40. In: Brock D, editor. Understanding Moore's Law: Four Decades of Innovation. Philadelphia, PA: Chemical Heritage Foundation; 2006. pp. 67-84

[2] Takahashi D. Forty Years of Moore's Law. San Jose, CA: Seattle Times; 2005

[3] Eigler DM, Schweizer EK. Positioning single atoms with a scanning tunnelling microscope. Nature. 1990; **344**:524-526

[4] Koenraad PM, Flatté ME. Single dopants in semiconductors. Nature Materials. 2011;**10**:91-100

[5] Lansbergen GP et al. Gate-induced quantum-confinement transition of a single dopant atom in a silicon FinFET. Nature Physics. 2008;**4**:656-661

[6] Calvet LE, Snyder JP, Wernsdorfer W. Excited-state spectroscopy of single Pt atoms in Si. Physical Review B. 2008; **78**:195309

[7] Tan KY et al. Transport spectroscopy of single phosphorus donors in a silicon nanoscale transistor. Nano Letters. 2010;**10**:11-15

[8] Hollenberg LCL et al. Charge-based quantum computing using single donors in semiconductors. Physical Review B. 2004;**69**:113301

[9] Schofield SR et al. Atomically precise placement of single dopants in Si. Physical Review Letters. 2003;**91**: 136104

[10] Fuechsle M, Miwa JA, Mahapatra S, Ryu H, Lee S, Warschkow O, et al. A single-atom transistor. Nature Nanotechnology. 2012;7:242-246

[11] Xie F-Q, Maul R, Wenzel W, Schn G, Obermair Ch, Schimmel Th. Single-atom transistors: Atomic-scale electronic devices in experiment and simulation. In: Proceedings of the International Beilstein Symposium on Functional Nanoscience; May 2010. pp. 213-228

[12] Fuechsle M, Miwa JA, Mahapatra S, Warschkow O, Hollenberg LCL, Simmons MY. Realisation of a single-atom transistor in silicon. Journal and Proceedings of the Royal Society of New South Wales. 2012;**145**(443 & 444): 66-74

[13] Obermair C, Xie F-Q, Schimmel T. The single-atom transistor: Perspectives for quantum electronics on the atomic-scale. Europhysics News. 2010;**41**:25-28

[14] Granot E. Exact model for single atom transistor. In: Stavrou VN, editor. Nonmagnetic and Magnetic Quantum Dots. Rijeka, Croatia: IntechOpen; 2018. Chapter 1

[15] Granot E, Azbel MY. Resonant angular dependence in a weak magnetic field. Journal of Physics: Condensed Matter. 1999;**11**:4031

[16] Granot E, Azbel MY. Resonant tunneling in two dimensions via an impurity. Physical Review B. 1994;**50**: 8868

[17] Granot E. Universal conductance reduction in a quantum wire. Europhysics Letters. 2004;**68**:860-866

[18] Miroshnichenko AE, Flach S, Kivshar YS. Fano resonances in nanoscale structures. Reviews of Modern Physics. 2010;**82**:2257-2298

[19] Azbel MY. Variable-range-hopping magnetoresistance. Physical Review B. 1991;**43**:2435

[20] Azbel MY. Quantum particle in a random potential: Implications of an

exact solution. Physical Review Letters. 1991;**67**:1787

[21] Granot E. Point scatterers and resonances in low number of dimensions. Physica E. 2006;**31**:13-16

[22] Abramowitz M, Stegun IA. Handbook of Mathematical Functions. New York: Dover Publications; 1972

[23] Granot E. Transmission coefficient for a point scatterer at specific energies is affected by the presence of the scatterer but independent of the scatterer's characteristics. Physical Review B. 2005;**71**:035407

[24] Granot E. Symmetry breaking and current patterns due to a weak imperfection. Physical Review B. 2000; **61**:11078

[25] Landauer R. Spatial variation of currents and fields due to localized scatterers in metallic conduction. IBM Journal of Research and Development. 1957;**1**:223-231

[26] Nazarov YV, Blanter YM. Quantum Transport: Introduction to Nanoscience. Cambridge, UK: Cambridge University Press; 2009. pp. 29-41